DIGITAL KIRLIAN PHOTOGRAPHY

Photographer's Guide to Shooting Spectacular Kirlian Photographs That Command Attention

by

John Iovine

Warning

By participating in the photography experiments contained in this book, you will be working near high voltage electricity, which can produce dangerous or lethal shocks. Please be careful and follow all directions and warnings contained within this book.

International Standard Book Number:

Print Book ISBN-13: 978-1-62385-004-3

Editor: Melissa Serao

Cover Concept: Greg Colletti

Photographs: John Iovine

Illustrations and Other Materials: Courtesy of the Author

Trademark Acknowledgements:

Table of Contents

Introduction

I have written on Kirlian photography for over two decades. This book focuses on using digital cameras to capture Kirlian images. My first book, "Kirlian Photography – A Hands on Guide" was written over 20 years ago, focused on using sheet film for creating Kirlian images.

In the intervening years since my first book's publication I occasionally dusted off my equipment, made a few modifications and took a few photographs. One major change that has occurred from when my first book was published and now is the advent of digital photography. In keeping up with the new digital equipment, I have written material detailing how to use digital cameras, iPhones and low light video cameras for capturing Kirlian photographs.

I have compiled this information into this book. In addition, I have added material for using photo-editing programs, like Adobe Photoshop, to reduce noise, enhance and colorize the resulting digital Kirlian photographs.

I also updated the Kirlian photography circuits and devices themselves. The new devices are more versatile and easier to build. You will find everything you need to create stunning Kirlian photographs in this book.

The Paranormal Claims

There are numerous Kirlian devices being sold on the internet today. To sell their devices, the corona discharge in the resulting Kirlian photograph is claimed to be paranormal or hold information that can be decrypted by the company's Kirlian researchers. All my research to date shows the corona discharge can be explained by employing known physical laws, like changes in subjects skin resistance (due to factors like; stress-lying, illness, fatigue, alcohol consumption, etc.) without the necessity of invoking paranormal (bio-plasma) factors. Some other variable factors influencing the resulting Kirlian picture, as described by Stanford researcher William Tiller, include the object's pressure against film, air humidity, air pressure, voltage, frequency, and exposure time. If anyone has any evidence to the contrary, I would be more than happy to examine that data.

My company also sells Kirlian equipment. However we have never made any "paranormal" claims to sell our Kirlian photography devices. Our devices are sold to make brilliant Kirlian photographs, period. If one wants to do paranormal research with our devices that fine. Just know we make no claim that any interpretation of the corona discharge is due to anything other than standard variable physical parameters as already previously outlined. If you discover something different, please let me know. I am open to examine any new data that comes along.

John Iovine 2016 NYC

Chapter 1

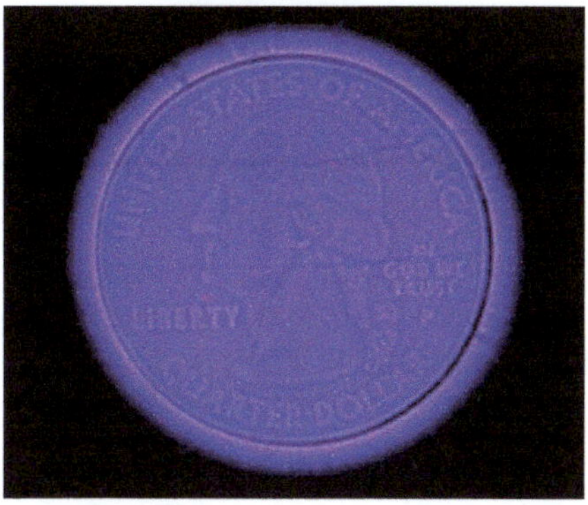

What is Kirlian Photography?

Kirlian (pronounced kur'-lee-uhn) *photography,* is sometimes called *electro-photography, electronography, electromagnetic discharge imaging (EDI),* and *radiation field photography.* It is commonly called *Kirlian photography* in honor of the Russian researchers Semyon and Valentina Kirlian, who pursued and developed this method of photography for over 30 years. Although they did not discover this method of electro-photography, they devoted a lifetime to studying its potential and made claims for its use as a diagnostic tool.

Kirlian photography uses high-voltage, low-current electricity to create a corona discharge around an object. The feeble light from the corona discharge can expose film or a CCD element in a digital camera. How the corona discharge is recorded using sheet film is different than when using a digital camera. So I will describe both methods, starting with classic film exposures.

Color Film Exposure:

When using color sheet film the object is usually in direct contact with the film, the corona discharge from the object exposes the film in two different ways. The first, is the feeble light from the corona discharge expose's the film in a normal standardized method for film exposure by light. The second, is when the corona discharge passes through the film, it activates the photosensitive silver halide crystals in the three dye layers of the color film. This is explained in better detail, later on in this chapter. Because of this dye activation, the resulting Kirlian photographed may contain colors that were not actually present in the corona discharge.

Figure 1-1 illustrates the basic Kirlian sheet film photography arrangement. The exposure plate is made out of a conductive material, usually a metal. The metal exposure plate is usually coated with a thin layer of insulating material, or dielectric, on its top surface. The sheet film lays on top of the exposure plate, emulsion side up, facing the object to be photographed. The object is placed on the film.

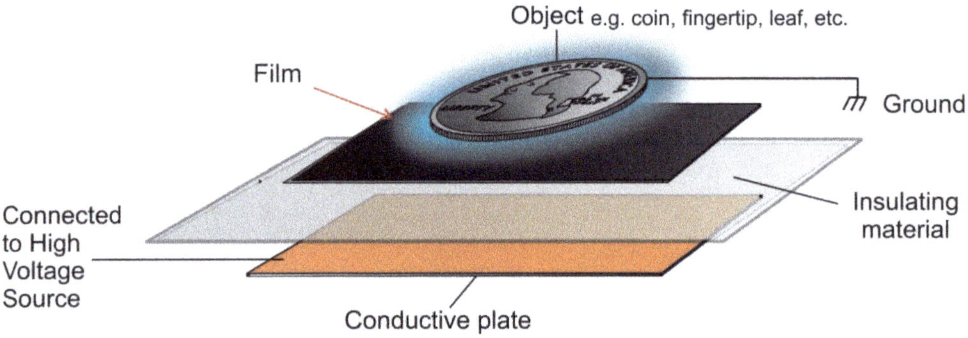

Figure 1-1 -- Overview of sheet film set-up

The high-voltage power source is connected to the plate. If the object is inanimate, it is connected to an earth ground. To take a Kirlian photograph, the room lights are turned off, or the Kirlian photography set-up is covered in a light tight enclosure. Next the high voltage power source is turned on, once on, a corona discharge forms around the object. The HV power is kept on for a short time, 2-3 seconds to exposure the film. After the film is exposed and developed, the resulting picture or transparency is a Kirlian photograph.

This simple illustration shows how Kirlian photographs can be made without light, lens system or a camera. This is Kirlian photography at its most basic level. It is a high voltage contact print photography.

Digital Camera, iPhone or Low Lux Video Camera:

This Kirlian photography set up may be used with the following types of cameras; digital, 35mm film, iPhone or a low lux video camera. I am illustrating the set up with a digital camera but the other cameras will work as well. The specifications for the cameras are provided in chapter 4.

Using a digital camera requires the use of a flat transparent discharge plate (TDP). The TDP itself is constructed using a glass plate with a transparent, but electrically conductive coating on one side of the glass. The transmission of light through the TDP is usually better than 90%.

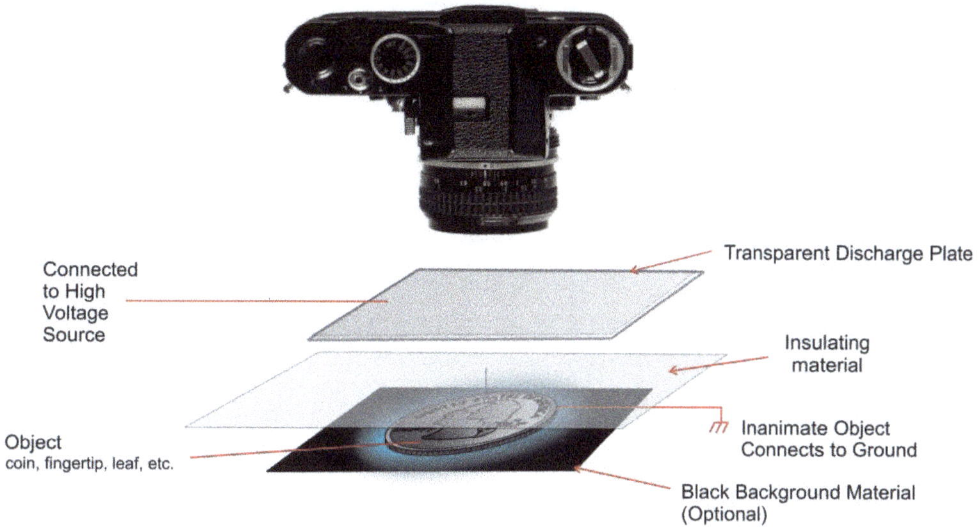

Figure 1-2 -- Overview of Digital photography set-up

The subject or object being photographed is in direct contact with the TDP. The high-voltage power source is connected to the TDP. If the object being photographed is inanimate, it is connected to an earth ground. The digital camera is manually focused onto the object through the TDP. The room lights are turned off, or the kirlian photography set-up is covered in a light tight enclosure. Next, the high voltage power source is turned on, once on, the feeble light from the corona discharge should be visible around the object. Next the digital camera shutter is tripped to make a timed exposure of 10-15 seconds. When the camera's shutter closes the high voltage power source is turned off. The digital camera should have recorded the corona discharge around the object which is a Kirlian photograph.

Directions: Art and Science

There are two main directions that can be taken in Kirlian photography. One is to use Kirlian photography for special effects photography. The other is to investigate the potential of this method of photography for its scientific value.

Aesthetics and Pictures

Using the Kirlian technique for special effects photography is easier than pursuing it for its scientific value. Many parameters that must be controlled and recorded for research are ignored when shooting for special effects. Aesthetically, the Kirlian technique can be used to achieve interesting special effects photographs. This book shows you how to build the equipment and the techniques to create a large variety of Kirlian photographs.

Scientific Inquiry

There was a flurry of research on Kirlian photography in the early 1970s. However, most of this experimentation was not under controlled laboratory conditions. This invalidated the work and conclusions drawn from it. In order for science to accept any data, the research and its results must be repeatable by other scientists. This is accomplished by using the scientific method with measured and controlled parameters.

Pursuing Kirlian photography for its potential scientific value is a harder road to travel. Any reports, of human bio-plasma fields and Kirlian auras that foretell illnesses will be met with a fair amount of skepticism from the scientific community. So, if you pursue scientific research, it's important to control every variable over every experiment and keep scrupulous records of controls and results.

We will not be pursuing Kirlian photography for its scientific value in this book. However, you can learn a little more about the science in Chapter B.

Photography variables

For the most part, as photographers we don't need to employ strict laboratory controls. However, we still want to produce the best pictures possible. Photographers learn that the patterns and "aura" of electro-photographs are predominately determined by conductivity of the object, pressure on the plate, air spacing between the object and the plate, length of exposure as well as moisture content of the air.

Color in Kirlian Film Photographs

Many Kirlian photographs show a wide range of colors, that includes reds, greens, blues, and white. If the corona discharge is predominately blue-violet, where do these other colors come from?

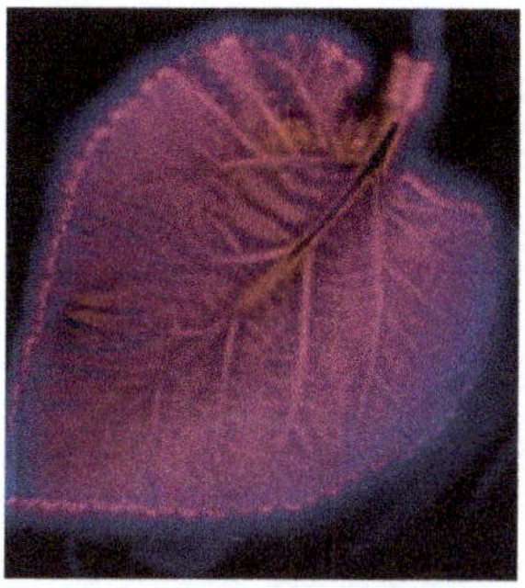

Figure 1-3 -- Kirlian image of leaf

To answer this question, you must examine the structure of color film. Color films are made with three separate silver halide dye layers. Each dye layer is designed to respond to a specific color of light: one for red, one for green, and one for blue (Figure 1-4). This arrangement works very well when exposing film to light in a standard camera. The colors in the image projected onto the film plane are accurately recorded in the film layers.

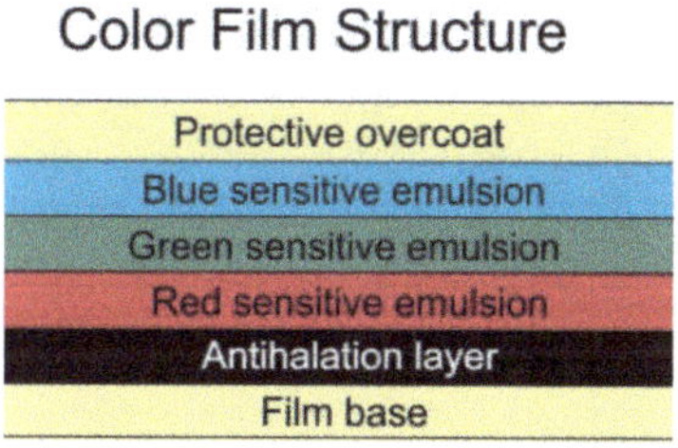

Figure 1-4 -- Structure of color film

With Kirlian photography, the high voltage electrical discharge (corona discharge) generates its light by photo ionization of the air, which is predominately nitrogen gas. This light contains a high component of UV light. If the film on the exposure plate buckles slightly, it creates an air gap where photo ionization can occur. Air gaps occur from too much pressure or uneven pressure between the object and the film. In addition, the corona discharge travels through the film layers where it can activate the silver halide crystals inside the dye layers and create color in the resulting Kirlian images that do not exist in the color of the corona discharge.

Color in Kirlian Digital Photographs

Digital cameras as well as other cameras; iPhone, video and film, that use a Transparent Discharge Plate, usually capture the true blue-violet color of the corona discharge in their photographs. However, the color of the corona discharge in the digital images may be changed, manipulated, and enhanced using a photo editing program like Adobe Photoshop (Figure 1-6). Photoshop may also be used to reduce and remove noise in digital photographs that typically accumulate in the long exposure times needed to capture Kirlian images as is shown in chapter 6.

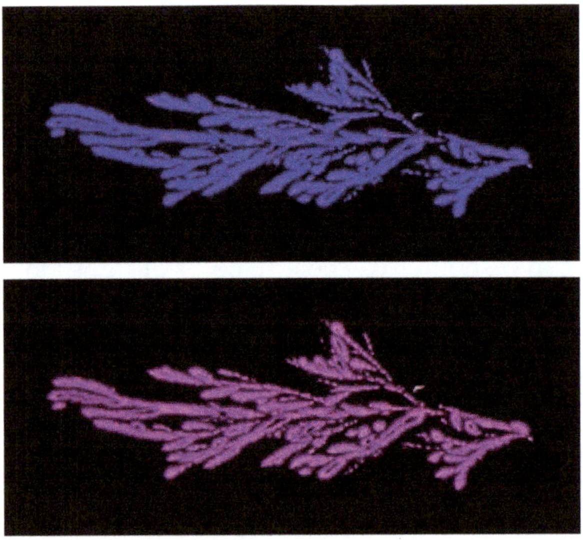

Figure 1-6 -- Kirilian images manipulated in Photoshop

Making Digital Kirlian Photographs

There are three main components you need to begin shooting digital Kirlian photographs, as well as a proper area or enclosure in which to shoot. A proper area is one where you can turn off the lights to achieve relative darkness. If a proper area is not available to you, you can instead, make a basic light tight enclosure, as detailed in chapter 5, for the Kirlian photography set-up. In each case darkness is the main requirement for shooting because it allows the camera to capture the feeble light from the corona discharge.

The three main components that are required are:

1) **High Voltage, low current power supply**

2) **Transparent Discharge Plate**

3) **Suitable Digital Camera**

You may buy or build a high voltage power supply and transparent discharge plate. In the next two chapters I will show you where you may purchase as well as how to build both. In chapter 4, I will go over the minimum requirements for your digital camera(s). The following chapters are on shooting Kirlian photographs and photo-editing the images for cleaning up noise and color manipulation.

Chapter 2

High Voltage Power Supplies

To create Kirlian photographs you need a high voltage (HV) power supply. You want a power supply that can supply a steady continuous output rather than one that outputs pulses. A reasonable high voltage power supply for Kirlian ought to be about to supply 3000 to 10,000 (or more) volts. However, the output current supplied by these power supplies I feel ought to be kept quite low. I point this out because; I have seen experimenters using what I consider unsafe high voltage high current power supplies.

What makes a high voltage power supply safe? Keep in mind that "Safe" is a relative term; all high voltage power supplies are potentially dangerous and will give a shock if used incorrectly. What I personally consider a safe power supply is one whose output current is 1 milliampere (1mA.) or less. I consider this low current to be non-lethal. And yes if you're interested I have shock myself many times when creating Kirlian photographs with my 1mA. HV power supply, while not pleasant, it's obviously non lethal. Voltage doesn't kill current does.

The trade off in using a low current HV power supply is the size of the object you can expect to use to create a Kirlian photograph. There is no hard and fast rule as to the size you can photograph with a 1mA HV supply, but from my experience you can expect to photograph an object with a 3" x 3" surface area with little problem. Keep in mind that the object's conductivity and topology factors heavily in the surface area one can expect to photograph.

Unsafe High Voltage Power Supplies

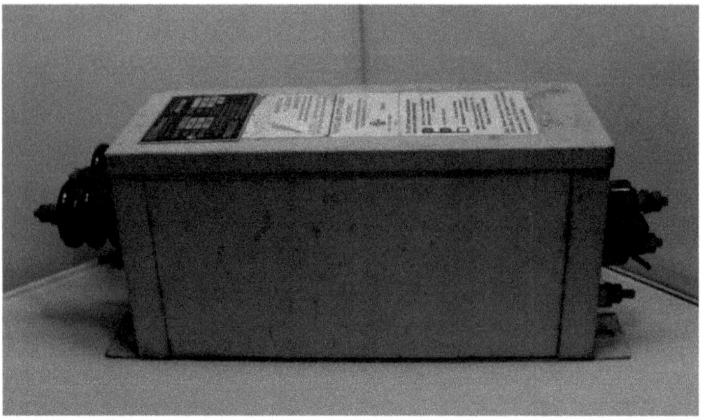

Figure 2-1 -- High voltage power supply (unsafe)

Before I move onto suitable power supplies for Kirlian photography I want to show you an example of an unsafe high voltage power supply. Neon Sign transformers are not safe, because they are high voltage high current. High current in these transformers varies between 20 to 30 milliamperes. I consider these transformers lethal. DO NOT USE!

Other high voltage circuits not to use are circuits/devices made for stun guns, Tesla coils, or any high voltage circuits/device that output more current than 1mA.

Safer High Voltage Power Supplies

There are a number of induction coil or Flyback transformer high voltage circuits that are suitable, low current. Induction coil circuits are typically lower frequency devices operating around 300 to 2000 Hz. Flyback transformer circuits typically operate at higher frequencies of 20000 Hz (20 KHz) or more.

The first question is to build or buy. There are a number of companies that sell high voltage power supplies and kits. I will look at one kit here. Now you do not have to purchase the kit. The schematic and parts list is also provided and you can use point to point wiring on a breadboard. The kit just makes it easier.

Important Safety Warning

This is a high voltage power supply that is intended for use by adults. Children should not build or operate this kit. This kit is not intended for children!

Assembly of this kit requires high-temperature soldering and the use of sharp edged components and cutting tools. Some included components may become hot, leak, or explode if used improperly. The author and Images SI Inc. recommend that you wear safety glasses when building or working with any electronic equipment.

High voltage discharges and shocks can cause injury and/or death. Additionally high voltage electricity as generated by this assembled kit can cause damage to property. The author and Images SI Inc. disclaim any liability for damage or injury caused by the use or misuse of the High Voltage Power Supply Kit. By building this circuit or kit, you agree not to hold the author or Images SI Inc. liable for any injury or damage related to the use, misuse or to the performance of this product. This product is not designed for, and should not be used in, applications where the malfunction of the product could cause injury or damage.

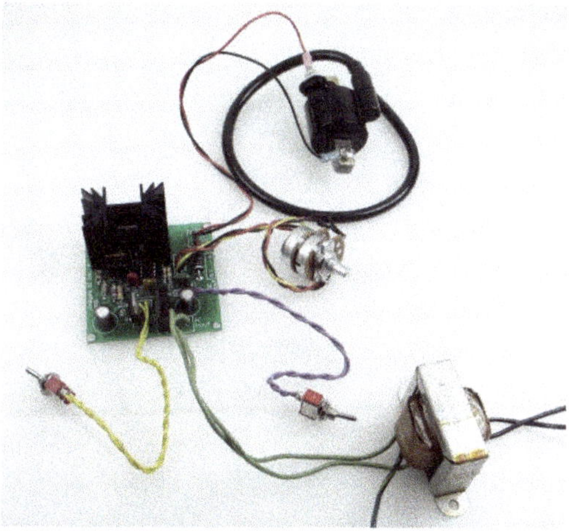

Figure 2-2 -- High Voltage Power Supply by Images SI, Inc.

Features:

This high voltage power supply features variable frequency control. It also has a High-Low frequency range switch. In addition, this circuit can be either battery powered 12 VDC, or powered from a wall transformer, 9-12

V with either an AC or DC output. Powering the unit from batteries provides the advantage of portability for fieldwork.

How the Circuit Works

The circuit schematic for the high voltage power supply is shown in the following figure. This is a simple device that is based on the 555 timer. The 555 Timer is set up as an astable oscillator. The output signal from the 555 Timer is fed to the base of Q3, a 3055 transistor. The signal from the collector of Q3 is fed to the base of a high voltage transistor Q1 that turns the transistor on and off, controlling the current flowing through the induction coil T1.

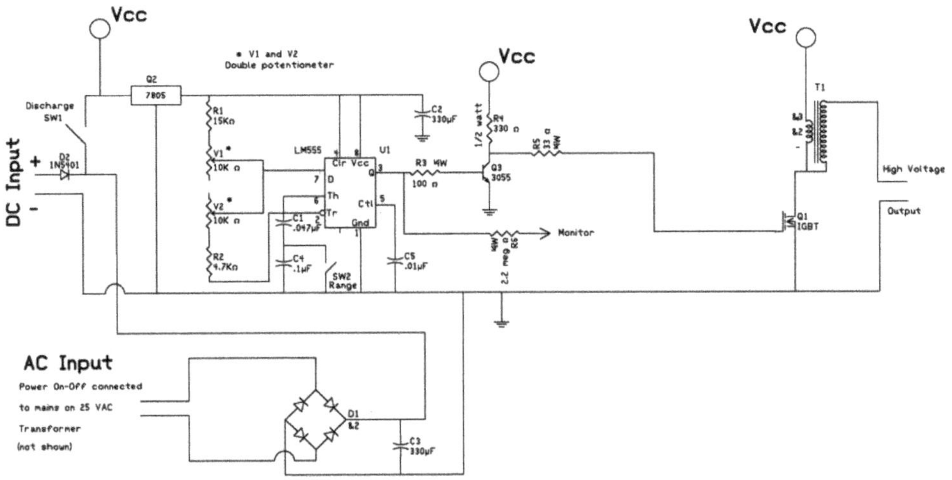

Figure 2-3 -- Schematic for high voltage power supply

The frequency of the 555 timer is controlled by the potentiometer and hi-low switch that adjust the timing capacitor. The potentiometer used in this circuit is a double-ganged potentiometer. Which means that it is two potentiometers that share a common shaft. A close up of the potentiometer is shown in Figure 5. The center terminal of the potentiometer is called the wiper. The two wipers of the potentiometers are soldered together and connected to pin 7 of the 555 Timer. All the current to energize the HV ignition coil passes through transistor Q1. To prevent Q1 from overheating, a large heat sink is attached.

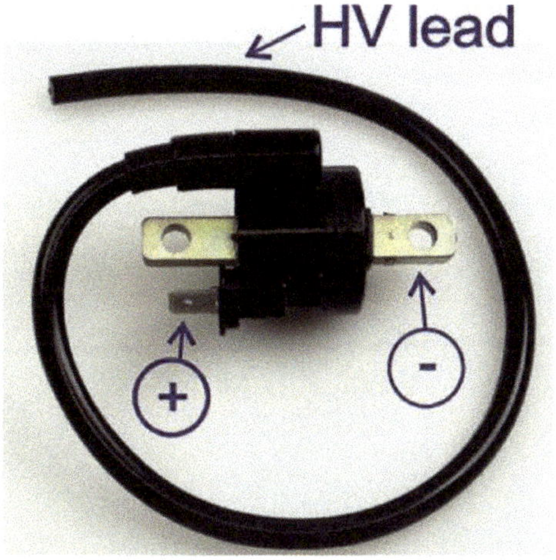

Figure 2-4 -- Ignition coil

The transformer T1 is a high voltage autotransformer. The T1 ignition coil transformer is a three terminal device. Figure 2-4 shows where we connect power from our circuit to the coil. The coil has a plastic protuberance that is the high voltage output of the transformer. A well-insulated (HV) wire is molded to this protuberance. The leads on the PC board labeled HV coil+ and – are used to connect T1 to pcb.

Construction of High Voltage Power Supply

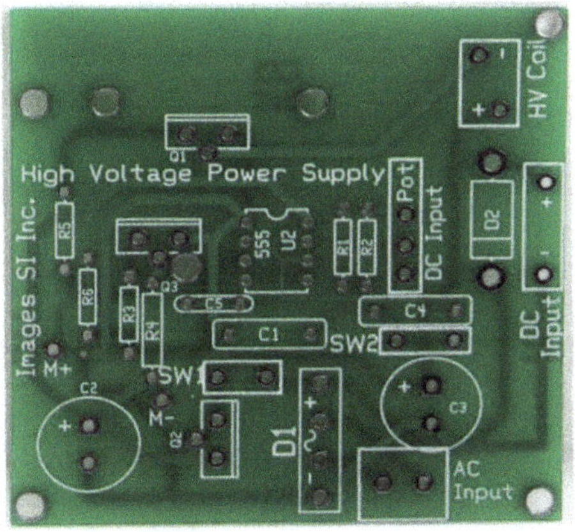

Figure 2-5 -- PCB

The components are mounted on the top side of the PC board. The top side of the board has white silk screen component drawings. The components are soldered on the opposite side of the pc board. After soldering the component to the board, any excess wire is clipped off.

Begin construction by mounting and soldering the 8-pin socket. Insert the IC socket, making sure to orient the notch on the socket to the drawing on the PCB and solder to the PC board.

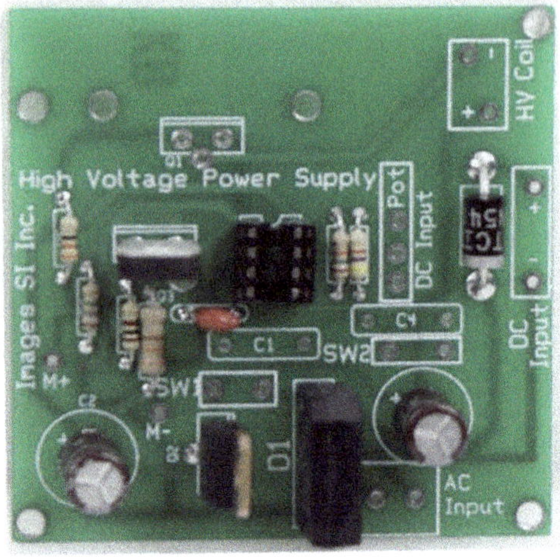

Figure 2-6 -- Construction of PCB

Next mount and solder Q2, the 7805 voltage regulator and Q3, the 3055. Mount and solder R1, the 15K (color bands brown, green, orange) resistor and R2, the 4.7K resistor (color bands yellow, violet and red). R3, the 100 ohm resistor (color bands brown, black, brown), R4, the 330 ohm 1/2 watt resistor (color bands orange, orange, brown), and R5, the 33 ohm resistor (color bands orange, orange, black) should now be mounted and soldered. Next mount and solder the bridge rectifier D1, making sure to orientated the + pin of the rectifier to the silk screen. Now, mount and solder capacitors C2 and C3, 330uF capacitors. Next mount and solder D2, the 1N5401 diode and C5, .01 uF capacitor.

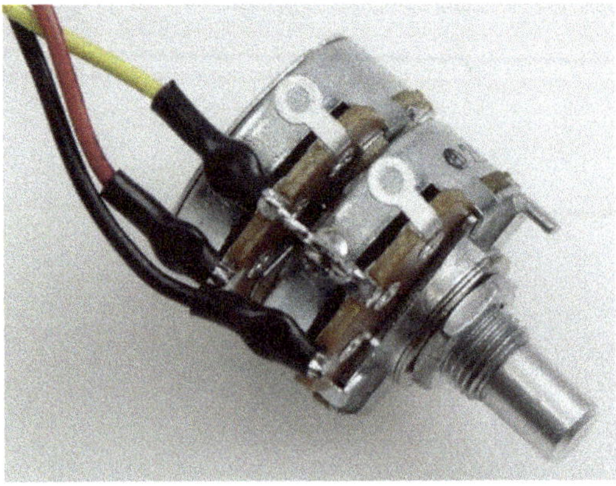

Figure 2-8 -- Double ganged potentiometer

Wire the double ganged potentiometer as shown in Figure 2-8. Next, solder the opposite ends of the wires to the printed circuit board, in the box labeled "POT". The center wire of the potentiometer is soldered into the center pad on the labeled box. The two end wires are soldered into the pads on either side of the center position.

Mount the Q1 transistor to the black aluminum heat sink. Next, mount the heat sink and transistor to the Q1 position on the PC board. The heat sink has two feet that fit into the holes on the PC board. Solder the heat sink feet in the holes to make a mechanically strong bond.

Attach 6 inch wires to both switches SW1 and SW2. Solder opposite ends of the wire into the pads marked SW1 and SW2 on the pc board respectively. Switch SW1 controls power to the circuit. Switch SW2 is the frequency high-low frequency control. Mount and solder capacitors C1 (.047uF), C4 (.1uF) to the PC board. Solder two 10" lengths of wire to the HV coil pads on the PC board. Attach the crimp terminals to the opposite ends on these wires. The quick disconnect terminal attaches to the (+) wire; the ring terminal to the (-). Attach these terminals to the HV coil as shown in Figure 2-9. Making sure to place the (+) lead to the (+) terminal on the HV coil.

Attach power leads to either the AC input or DC input pads on the PC board.

Next install the integrated circuit. When installing integrated circuit (IC) chips, begin by first identifying the top of the chip. The top of the chip has a marker, many times it is a half circle cutout. Sometimes it is a small mark identifying pin 1 on the IC. In both cases the marks show us the top of the IC chip. Orientate the top of IC chips with the white silk screen drawings of the components on the top of the pc board (usually a half circle cutout) or on the parts placement drawings and install the IC into their socket. The leads on the PC board labeled M+ and M– are not used.

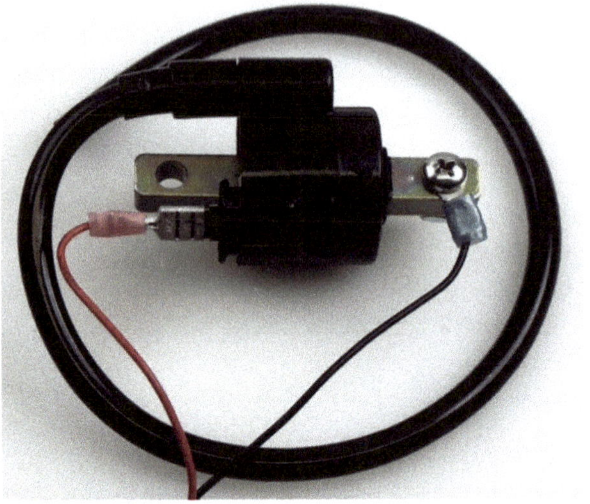

Figure 2-9

Testing & Finishing Construction

To test the circuit, take the open end of the HV wire and place it about ¼" away from the (-) terminal on the ignition coil. Apply power to the circuit. An electrical discharge should jump between the HV wire and the negative terminal of the ignition coil. Adjust the potentiometer (frequency control) to obtain the largest spark across the discharge.

If you do not get a continuous HV spark, you have a board error. Go back to your pc board and start checking your components and soldering.

The working circuit should be mounted inside a plastic enclosure. Coat any exposed wires with a plastic spray to provide insulation (No-Arc spray is available at your local Radio-Shack store. Corona dope is another insulating material. In a pinch you can use clear nail polish. Since nail polish is flammable, allow the nail polish to completely dry before using the circuit.)

Frequency & Voltage Output

Using the standard C1 and C4 capacitors the low frequency range is approximately 600-1300 Hz. The high frequency range is approximately 800-2000 Hz. By changing capacitors C1 and C4, one can vary the operating frequency and output voltage of the circuit, see chart below.

NOTE: Using a 24 VDC input voltage to the circuit, the voltage output, measured using a spark gap, is approximately 10-25 kV. When using a 24V transformer, the output is so high that the voltage arcs across the transformer. It is recommended that you completely encase the high voltage transformer in oil to prevent arcing.

HV TRANSFORMER IN AIR INPUT VOLTAGE 12 VDC

C1(uF)	C4(uF)	Low Frequency Range	Output Voltage	High Frequency Range	Output Voltage
.047	.1	583 Hz / 1268 Hz	5 / 12 kV	862 Hz / 1874 Hz	3 / 10 kV*
.22	.22	123 Hz / 270 Hz	10 / 12 kV	250 Hz / 244 Hz	12 / 15 kV
.22	.47	123 Hz / 268 Hz	12 / 15 kV	181 Hz / 395 Hz	12 / 14 kV
1.0	1.0	29 Hz / 62 Hz	7 / 10 kV	56 Hz / 122 Hz	8 / 15 kV

HVPS-01 Parts List

	PCB-36
	Heatsink-03
C1	**.047 uF, 100V capacitor**
C2,C3	**330 uF, 16V capacitors**
C4	**.1 uF, 100V capacitor**
C5	**.01 uF, 100V capacitor**
Q1	**HV Transistor IRF830 or Equivalent**
Q2	**7805 Voltage Regulator**
Q3	**3055 NPN Transistor T0220**
U2	**LM555 Timer**
	8 pin socket
T1	**HV transformer (HVT-09)**
R1	**15K 1/4 watt resistor**
R2	**4.7 K 1/4 watt resistor**
R3	**100 ohm 1/4 watt resistor**
R4	**330 ohm 1/2 watt resistor**
R5	**33 ohm 1/4 watt resistor**
R6	**2.2 Mega ohm 1/4 watt resistor**
D1	**Bridge rectifier (BR-4A-100)**
D2	**1N5401 diode**
SW1,SW2	**SPST toggle switches(SW18)**
POT	**Double ganged 10K potentiometer**

(1) Screw 540x3/8 PH Z
(1) Nut 540x5/16 SS
(1) Ring Terminal (blue)-Terminal-34161
(1) 1/4" Quick Disconnect Terminal (red)
(1) Screw 1024x5/8-PH
(1) Nut 1024x3/8Z
(1) Lockwasher-LW-10

Enclosure

You have completed the HV circuit and tested it. You can use the circuit as is, but it is a better idea to enclose the circuit inside a plastic case. I recommend a plastic case, to help eliminate shorts and shocks.

The size of the case I used is 6”x 6” x 3.25” deep. Any size case large enough to hold the circuit may be used. I created a decal for the top of the enclosure that you can download and print for free from the Imagesco website.

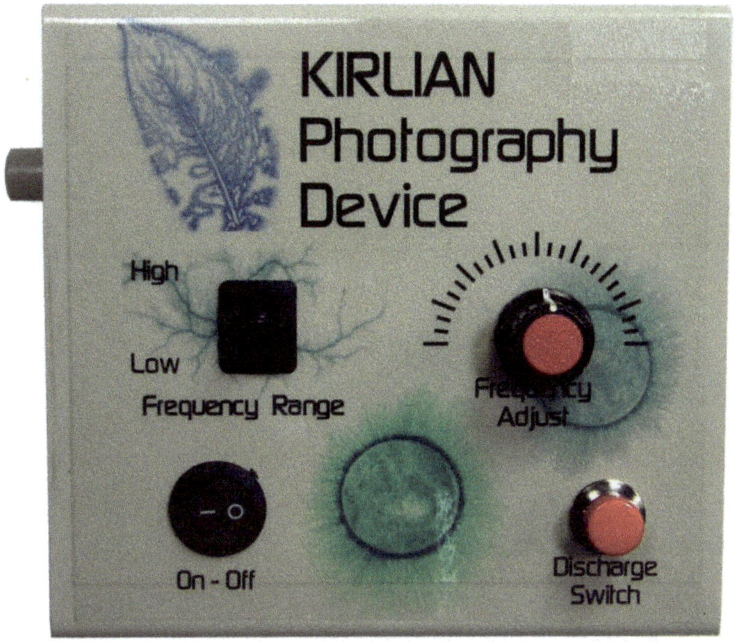

Figure 2-10 -- Top View of Kirlian Enclosure

I improved the standard circuit by adding an On-Off power switch that controls power to the unit. I upgraded the Discharge switch from a standard on-off toggle switch and made it a Normally Open (OFF) momentary contact switch. To activate the HV power supply (turn on) both the On-Off power switch must be On AND the Discharge switch must be pressed.

Using a momentary contact switch makes it easier to time exposures.

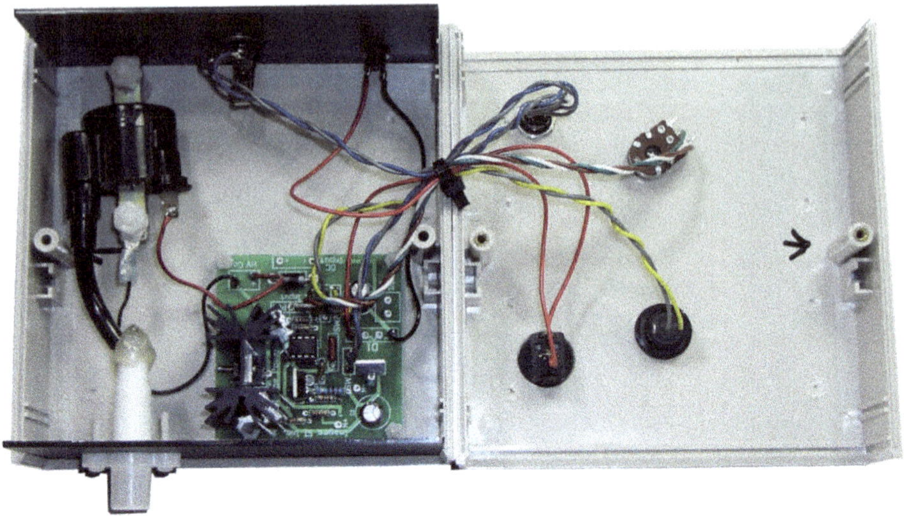

Figure 2-11 -- Internal View

The photograph above shows how I placed the circuit board, HV transformer and high voltage socket inside the plastic enclosure. Notice that the 12V power transformer is not mounted inside the enclosure. I decided to use a 12 volt wall transformer. The transformer has a 2.5mm power plug. So I mounted a 2.5mm power socket inside the enclosure to connect the external power to.

Handling high voltage is always a concern. So to make connecting to the high voltage power supply safer I decided to use connectors specifically manufactured for high voltage see figure below.

HV Connectors

Below are high voltage connectors, a socket and a plug.

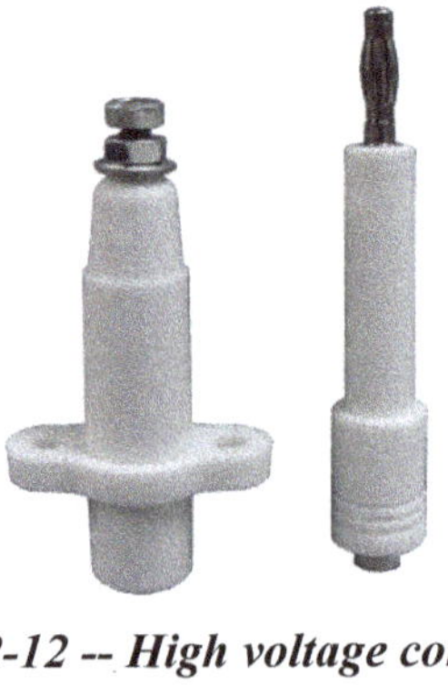

Figure 2-12 -- High voltage connectors

I connected the high voltage output from the HV power supply to a HV socket I mounted on the wall of the plastic enclosure. Notice I used non-conductive plastic screws and nuts to mount the HV socket. The reason was to not give the HV power a conductive object to possible spark and jump to. See picture below. After I attached the HV wire to the socket, I coated the metal screw and nut of the HV socket with hot glue as an added measure of insulation and to prevent the nut from getting loose.

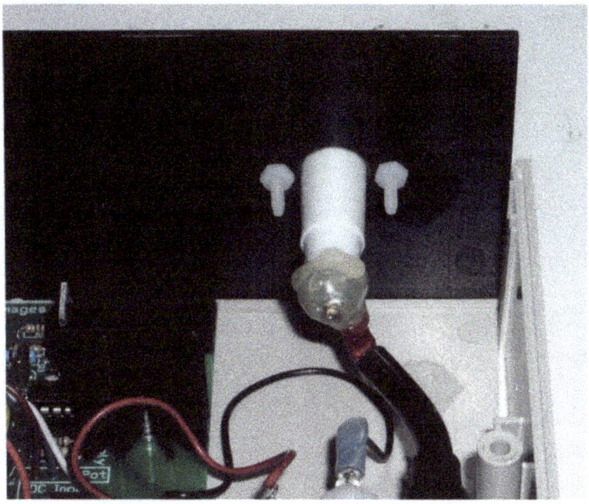

Figure 2-13 -- Mounting connector inside enclosure

From the outside of the enclosure the connector socket looks like this:

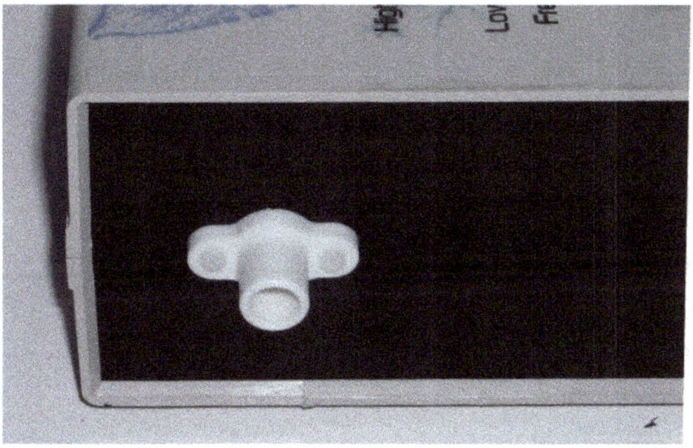

Figure 2-14 -- View of connector from outside case

To connect anything to the high voltage power suppy I used the high voltage plug. So the high voltage wire from the Transparent Discharge Plate (TDP) is connected to a high voltage plug, which in turn is plugged into the socket.

The advantage of this is that you can connect other accessories to the HV power supply. For instance if you want to try taking Kirlian photographs using B/W or color sheet film, you could use the Sheet Film Plate (SFP) instead of the TDP.

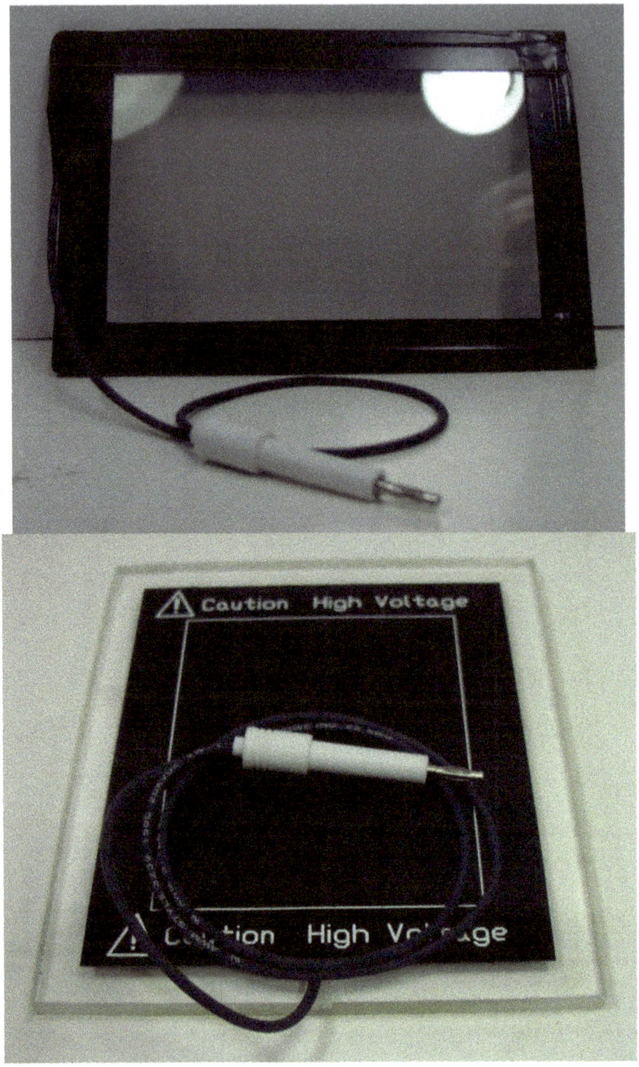

Figure 2-15 -- Transparent Discharge Plate & Sheet Film Plate

Chapter 3

Transparent Discharge Plate

The Kirlian photograph of the U.S. quarter below was shot with a digital camera through a transparent discharge plate (TDP).

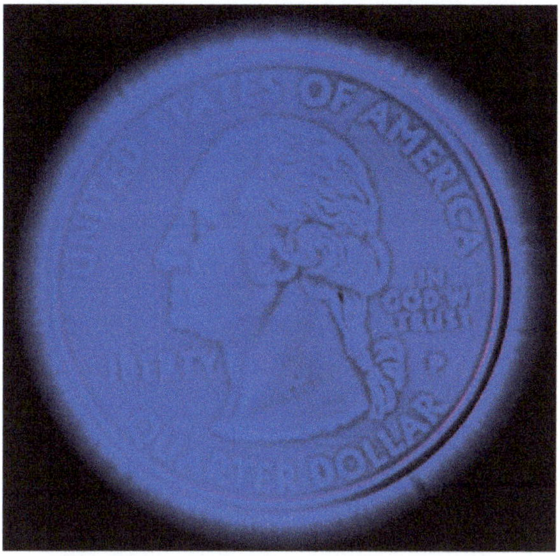

Figure 3-1 -- Kirlian Image of Quarter

Transparent Discharge Plate - Build or Buy

Your first decision is whether you should build or buy a Transparent Discharge Plate (TDP). To build a TDP will save you about half the cost of buying one. The retail price of a small TDP is $100.00 and building one will cost about $50.00.

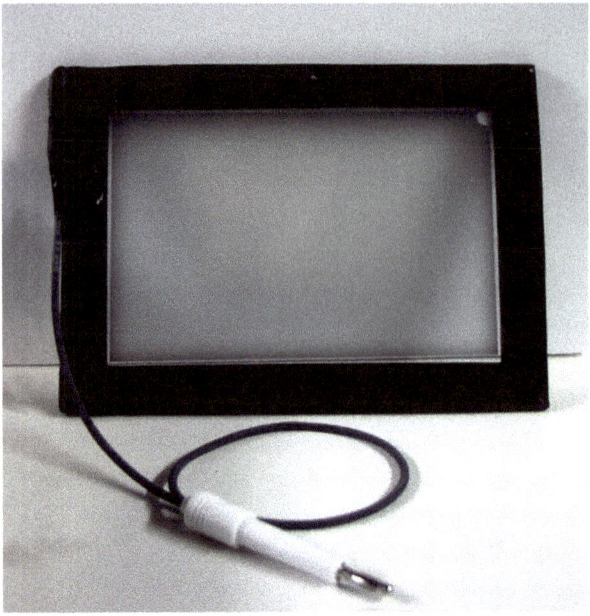

Figure 3-2 Transparent Discharge Plate from Images SI Inc.

To purchase a TDP go to:
http://www.imagesco.com/kirlian/transparent_discharge_plate.html

The balance of this chapter details the construction of the Transparent Discharge Plate and a Ground Plate. A ground plate isn't necessary; one can connect a ground wire when it's appropriate to do so. However, a ground plate is very simple to make and will make your photography set-ups easier to do and improve your Kirlian images at the same time.

Construction Of A Transparent Electrode

A transparent discharge plate is constructed from four main components; a 4 inch by 5 inch piece of glass that has a transparent, electrically conductive coating of tin oxide on one side, a ring electrode made from thin adhesive copper foil, and two .005 thick pieces of transparent plastic. The transparency of our transparent discharge plates surpasses 90 percent.

The conductive tin oxide coating on the glass can be damaged by a high voltage discharge. To prevent damage a solid electrical connection must be made between the tin oxide coating on the glass and the high voltage power source. The solution for this is simple, we make a ring electrode using conductive copper tape along the perimeter of the glass on the tin oxide side. The adhesive copper tape creates a ring electrode with a large surface area of electrical contact with the tin oxide coating. This large surface area prevents a discharge (spark) forming between the electrical connection to the tin oxide coating and the tin oxide coating itself, when in use.

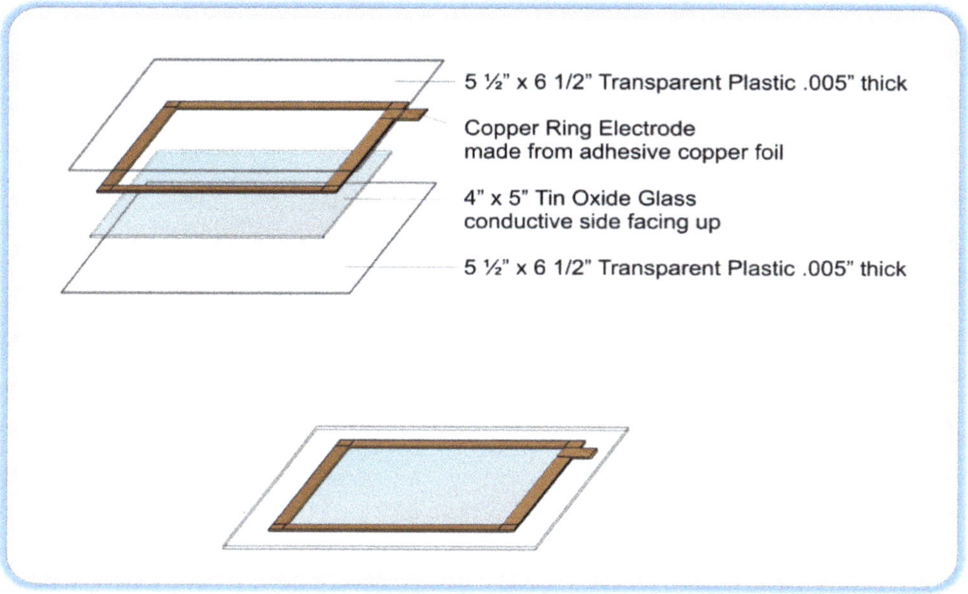

Figure 3-3 – Overview, The Components of Transparent Electrode

The transparent discharge plate construction is shown in figure 2. The rectangle ring electrode is made from thin adhesive copper foil, that is 1/4 inch wide and 2 mils thick. You will need two .005 thick pieces of transparent plastic sheet material. The thin transparent plastic sheets are cut 1-1/2" larger than the glass plate on each side, making it approximately 5-1/2" by 6-1/2", as shown in Figure 2 above.

When you receive the 4" by 5" tin oxide glass (Figure 3), be careful. The edges of the glass are sharp. Visually it will be impossible to tell which side of the glass has the conductive tin oxide coating.

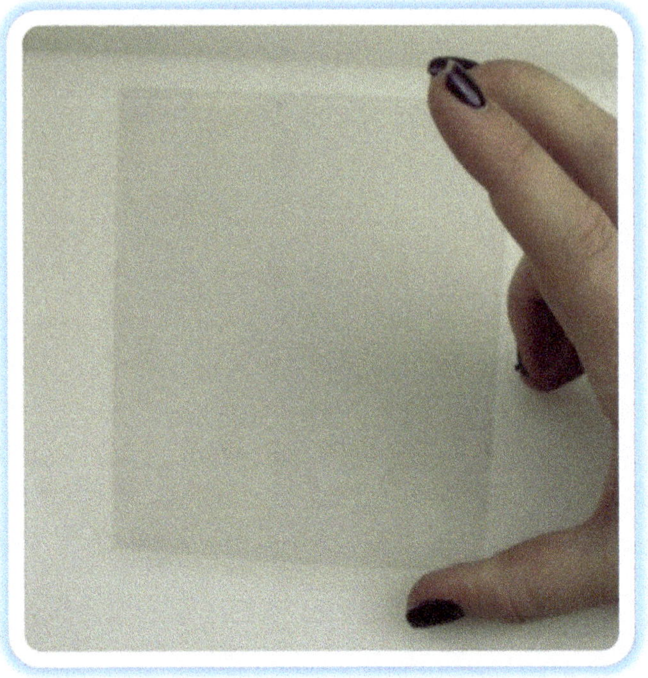

Figure 3-4 - Conductive Glass

We begin by determining which side of the glass has the tin oxide coating. To do this you need an ohmmeter. Inexpensive VOM's (volt-ohm-meters) are available from local electrical stores or from Amazon.Com. An inexpensive VOM can be purchased for $10.00.

http://www.amazon.com/Volmate-Digital-Voltmeter-Ohmmeter-Multimeter/dp/B00TWTLWT8/ref=sr_1_1?ie=UTF8&qid=1464961279&sr=8-1&keywords=VOM

Set the VOM to read OHMS on lowest resistance setting. Bring both probes in contact with one side of the glass plate. The side with the tin oxide coating will deflect the meter, or if you are using a digital VOM, as shown in figure 4, the resistance reading is provided numerically. The VOM in figure 4 is showing a resistance of 165 ohms per inch. The other side of the glass, the side without the tin oxide coating will show an infinitely high resistance.

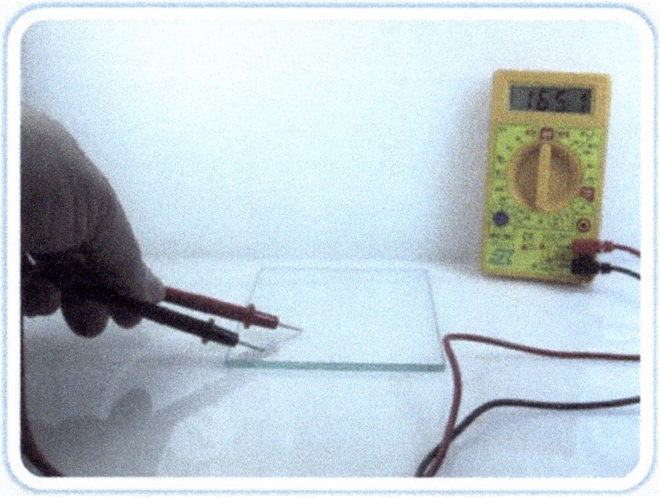

Figure 3-5 - Testing for Conductivity

The thin adhesive copper foil can easily be cut using scissors. Before placing the copper foil on glass, clean the tin oxide side of the glass thoroughly, so the foil has a good surface to adhere to. Re-clean the glass surfaces throughout the transparent discharge plate's assembly to ensure that no prints or lint is left on the glass.

The outside dimension of the ring electrode is the same size as the glass plate. Cut two 4 inch strips and two 5 inch strips of copper foil. These strips of foil are equal to the length and width of the glass plate. Remove the adhesive backing and place foil along the edge of the conductive side of the glass, as shown in Figure 5. After the copper foil has been secured to the four edges of the glass, cut a one inch piece of copper foil and adhere it the lower right side of the glass's copper ring electrode. Fold any excess foil in half to create the tab, see Figure 6. This copper tab is where we will solder our high voltage wire to.

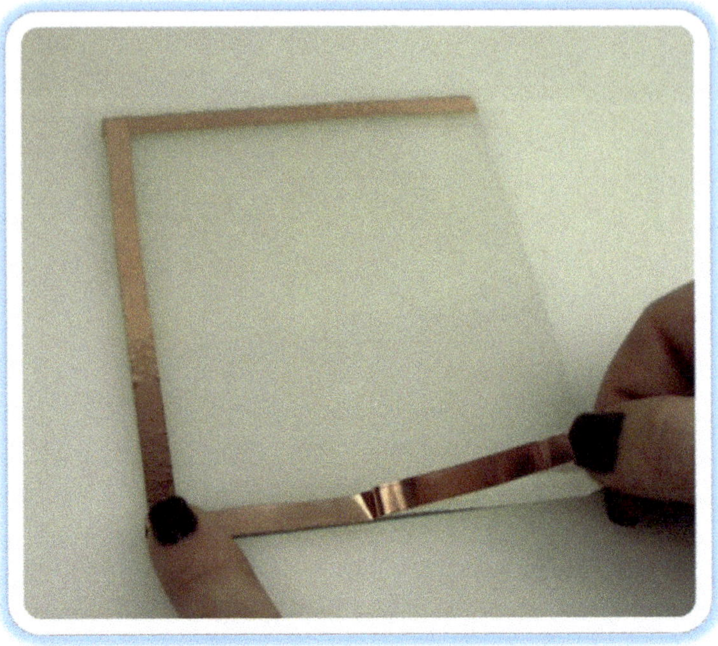

Figure3-6 - Laying Copper Foil

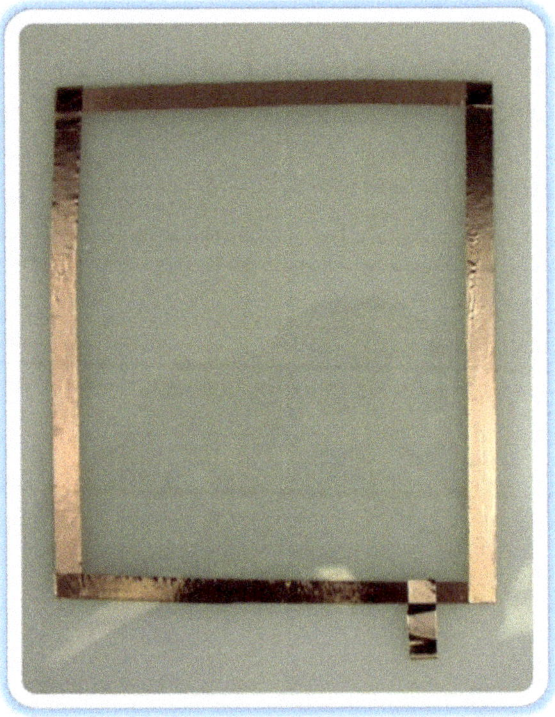

Figure 3-7 - Completed Electrode Ring with Tab

Test assembly with an ohmmeter before proceeding. Touch the tab of the ring electrode with one probe and touch the tin oxide coating with the other. You should show good conductivity (low resistance).

Next, solder approximately 18 inches of high voltage wire to the protruding tab. The wire should lay perpendicular to the copper tab as shown in Figure 7. Try to solder the wire so that it lays as close to the glass as possible. This will connect the transparent discharge plate to the HV source when you are finished.

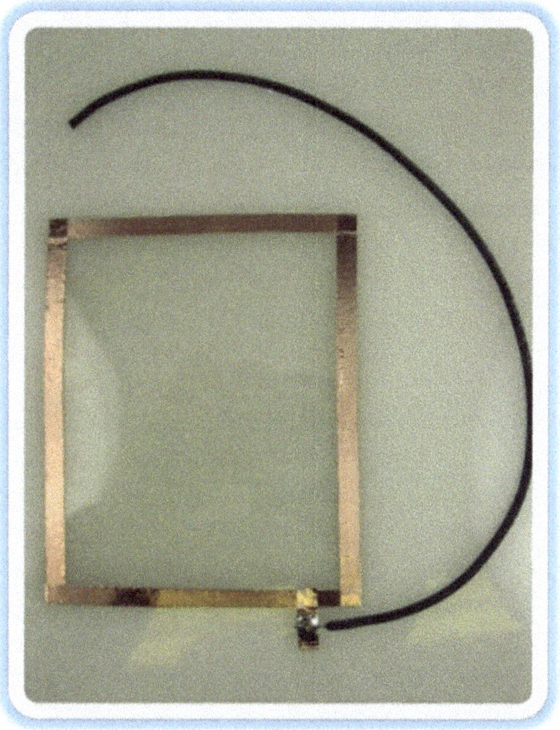

Figure3-8 - Electrode with 18 inches of High Voltage Wire Soldered to Tab

Since the conductive coating of the transparent electrode is delicate we need to protect it. Purchase two sheets of transparent plastic which is approximately .005 thick. Cut the plastic sheet to a larger size than the tin oxide glass. Add approximately 3/4 inch to all sides of the transparent plastic.

Run a thin bead of clear silicone adhesive around the edge of the transparent electrode, on top of copper foil, as shown in Figure 8. Center the plastic on top of the conductive tin oxide coating (Figure 9). Place a book or flat object onto the transparent electrode to hold the assembly tight and together while the silicone cures. Curing time is approximately 12 - 24 hours.

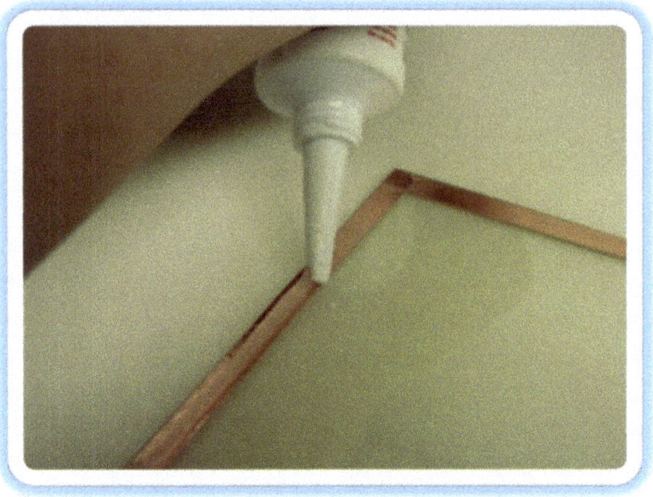

Figure 3-9 - Running Bead of Silicone

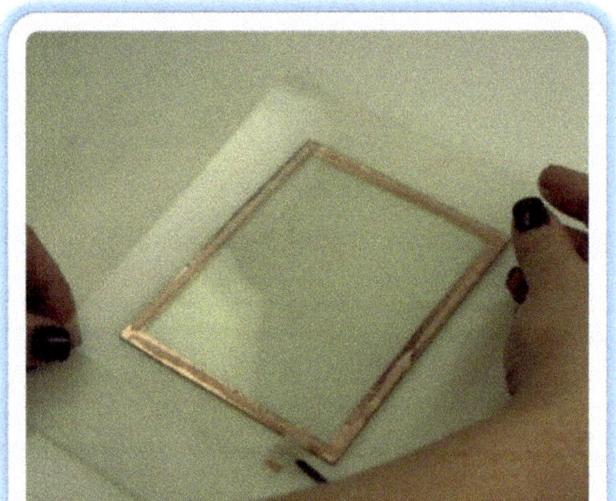

Figure 3-10 - Center Acetate over Electrode

Once the silicone is dry, repeat this process on the opposite side of glass by running a bead of silicone around the edge of glass and centering second piece of plastic on top of glass, as demonstrated in Figures 10 and 11. Place a book or flat object onto the transparent electrode to hold the assembly tight and together while the silicone cures. Wait another 12 -24 hours.

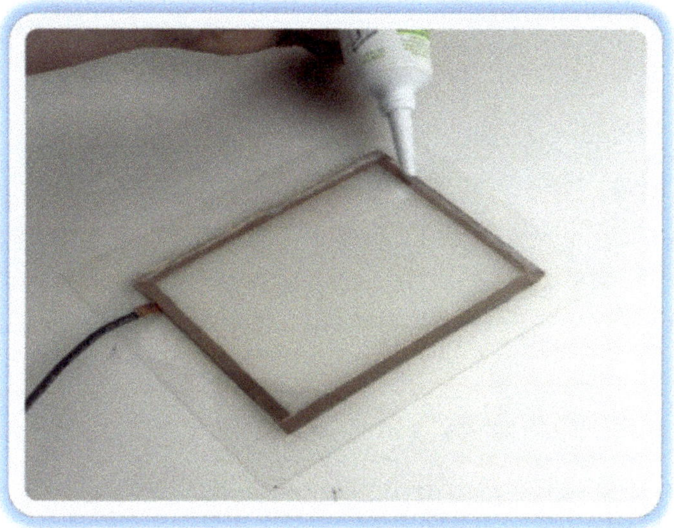

Figure 3-11 - Run Bead of Silicone on Glass Side

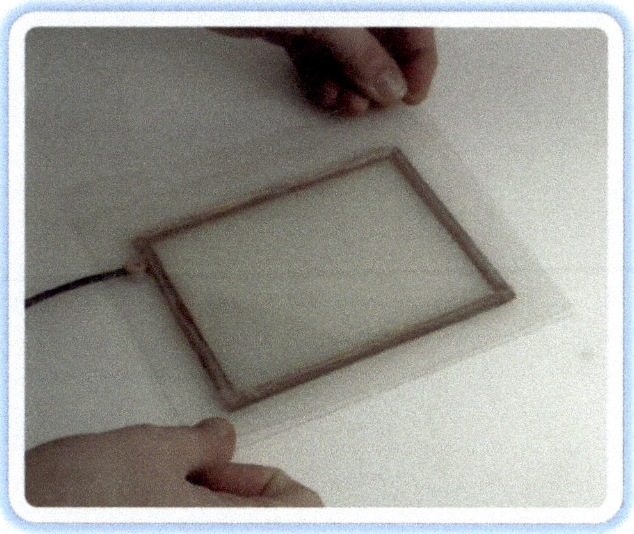

Figure 3-12 - Center Second Sheet of Acetate

You should now have the transparent electrode sandwiched between two sheets of transparent plastic. Next, fill the gap from the edge of the glass to the outside edge of the plastic sheets with silicone adhesive, see figure 12. Place a book or flat object onto the transparent electrode once again to hold the assembly tight and together while the silicone cures. Allow to dry completely before continuing.

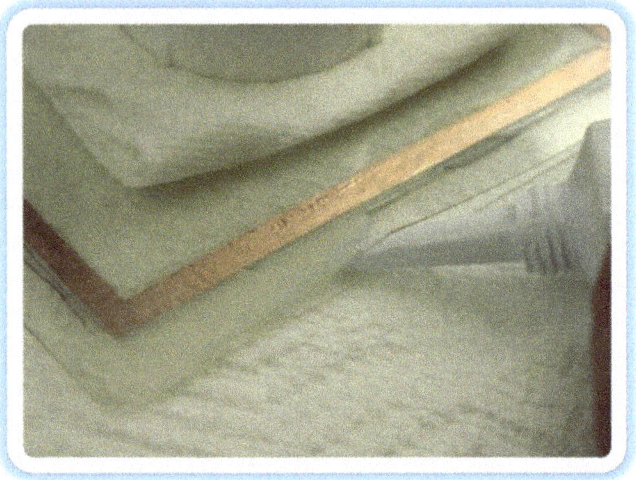

Figure 3-13 - Fill Gap Between Acetate Sheets with Silicone

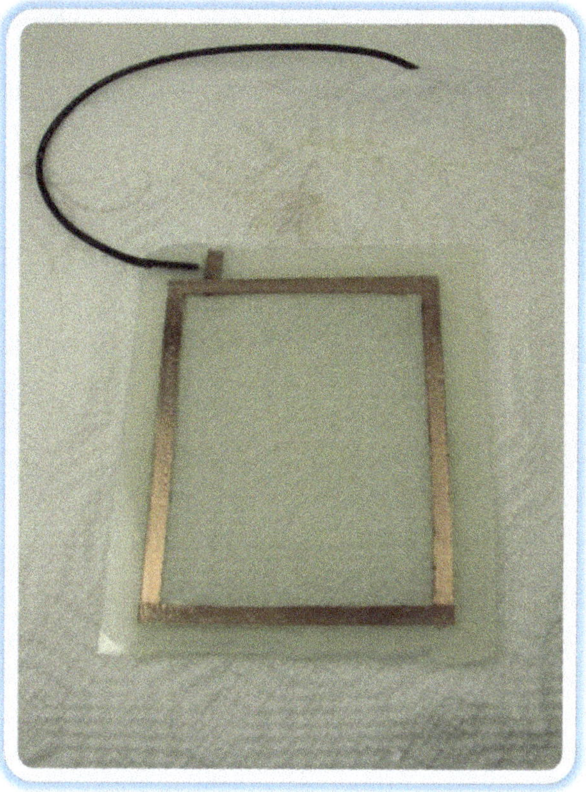

Figure 3-14 - Electrode Sealed in Silicone and Acetate

Create a frame for your transparent electrode by running black electrical tape along the edges of the transparent electrode to further secure the plastic sheets together (see Figure 3-15).

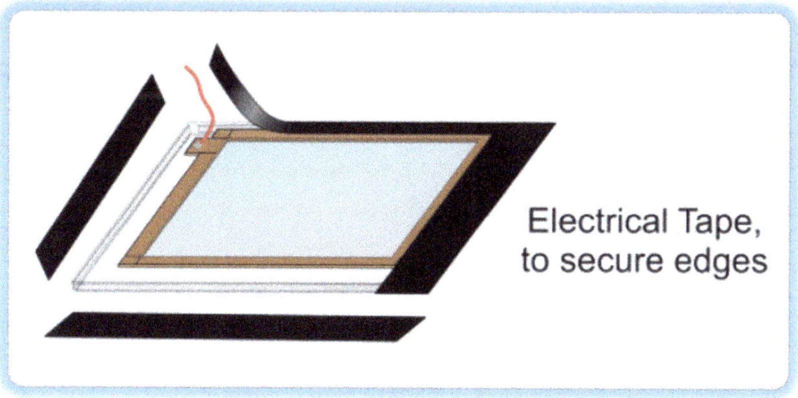

Figure 3-15 - Framing Electrode

First, run the tape along the edges of the transparent electrode. When you reach the edge that has the HV wire protruding from the side, use an Exacto knife to cut a slice in the tape, so that the tape can be secured to electrode without leaving a gap. Figures 14 -18 demonstrate this process.

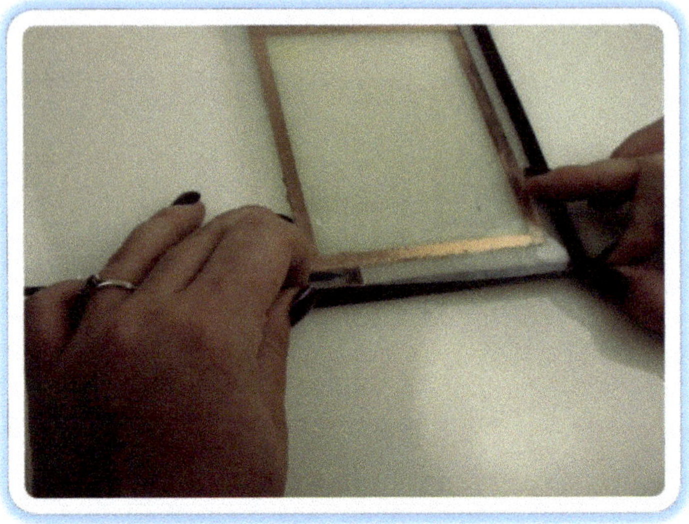

Figure 3-16 - Taping Edges of Electrode

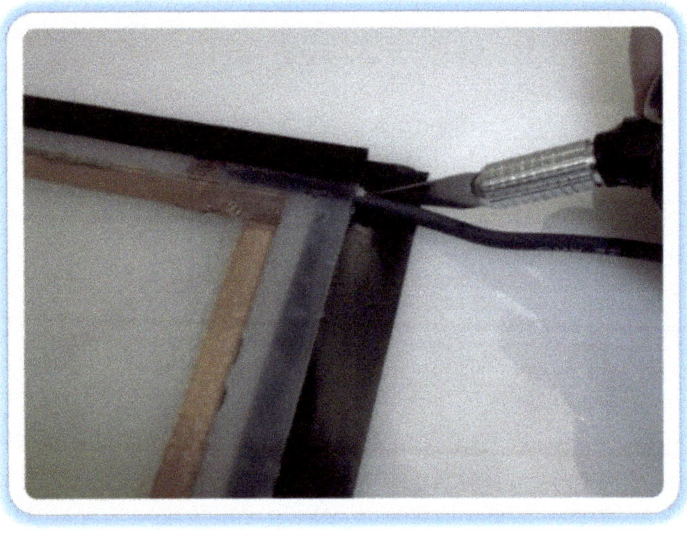

Figure 3-17 - Cut Slit for Wire

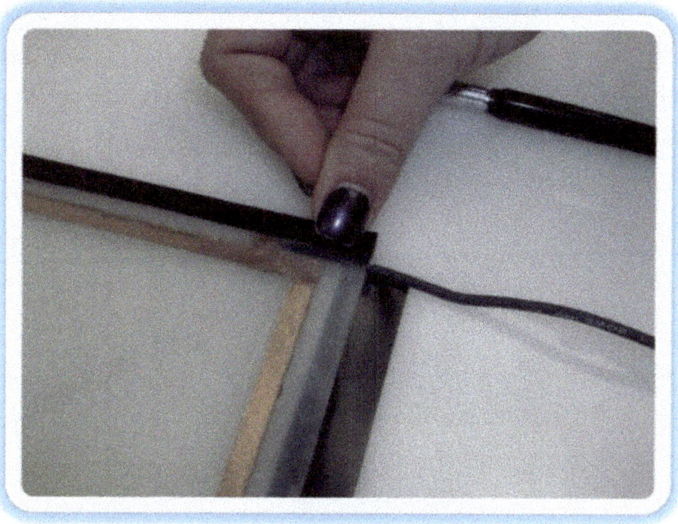

Figure 3-18 - Fold up Small Piece of Tape

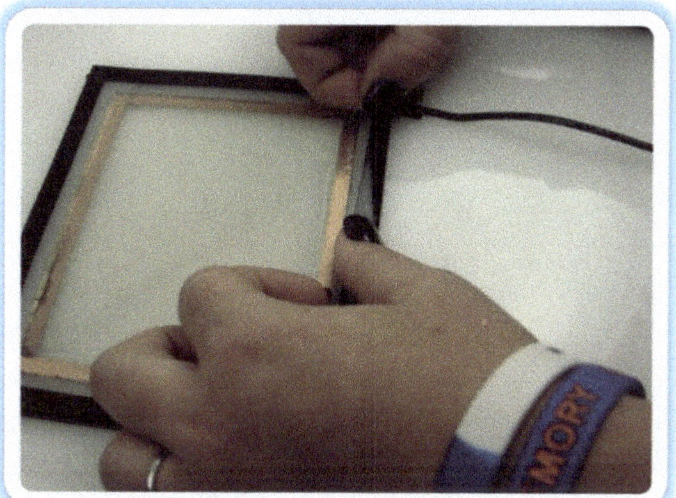

Figure 3-19 - Finish Taping Along Edge

Then, wrap it around the glass on each side to cover the space between the edge of the electrode and where the copper foil appears, as demonstrated in Figures 3-20 and 3-21.

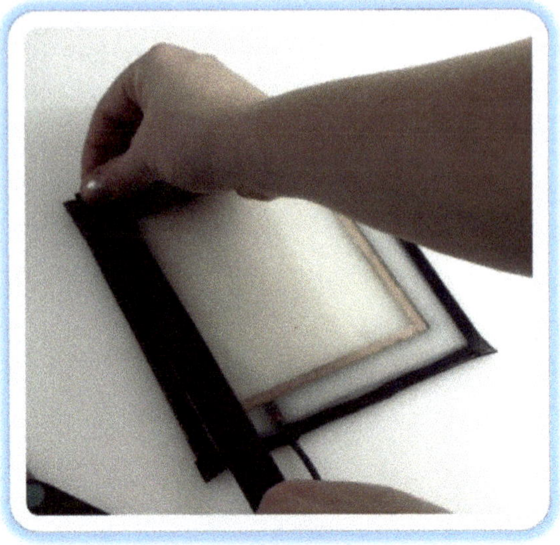

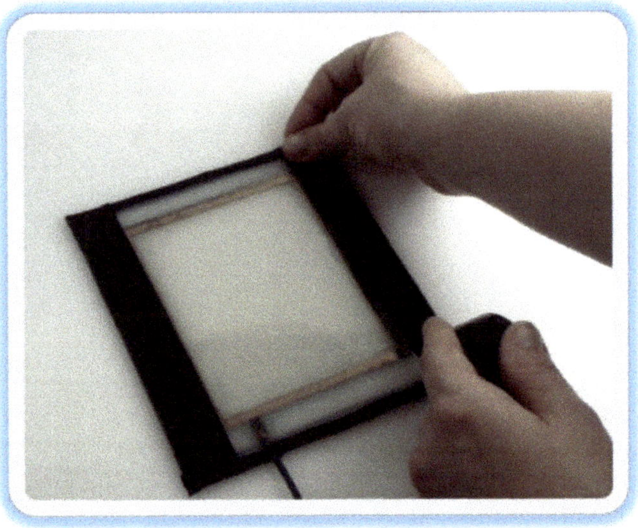

Figures 3-20 & 3-21 - Wrap Tape Around Glass to Cover from Edge to Copper

The tape serves a dual purpose. First, it helps secure the plastic sheets to the transparent electrode. Secondly, it provides insulation along the edge of the electrode where it is needed. You may want to double tape the corner edges to provide an additional measure of insulation.

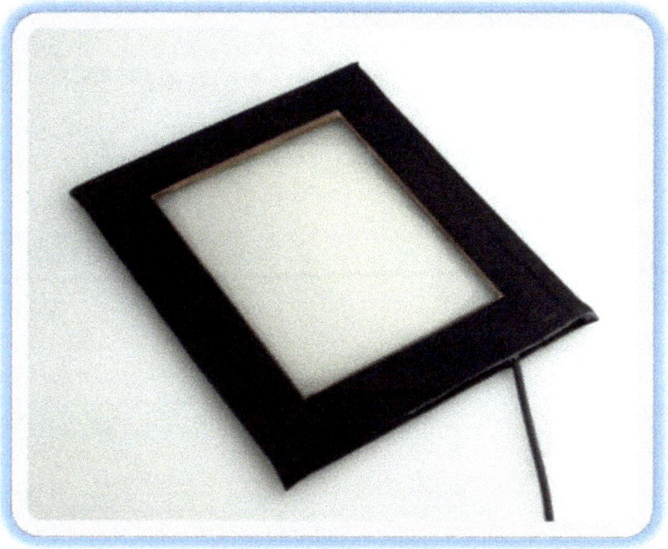

Figure 3-22 - Finished 4" x 5" Transparent Discharge Plate Ready for Use

Parts List:

Tin Oxide Coated (1 side only) Glass 4" x 5"

¼" wide copper adhesive tape. Approximately 24" in length

Clear silicone caulking/glue

(2) 5-1/2" x 7" clear transparent plastic 0.050 thick

Length of HV wire (8-10Kev) approximately 18"

Black electrical tape

Components and Transparent Discharge Kit available from Images SI Inc.

http://www.imagesco.com/kirlian/kirlian-components.html

Purchasing a Transparent Discharge Plate

If you prefer to buy rather than build, you can purchase a transparent discharge in a number of sizes from Images SI Inc.

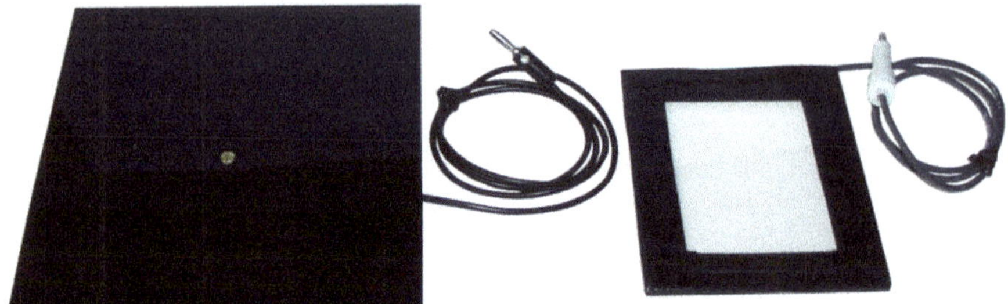

The Images Transparent Discharge Plate (TDP) includes a ground plate.

http://www.imagesco.com/kirlian/transparent_discharge_plate.html

Ground Plates are very useful and are explained in the next section.

Ground Plate

Being about to easily ground the object you want to photography safely and quickly is essential. I recommend making a few grounding plate tools to facilitate your Kirlian photography.

Ground Plate (GP)

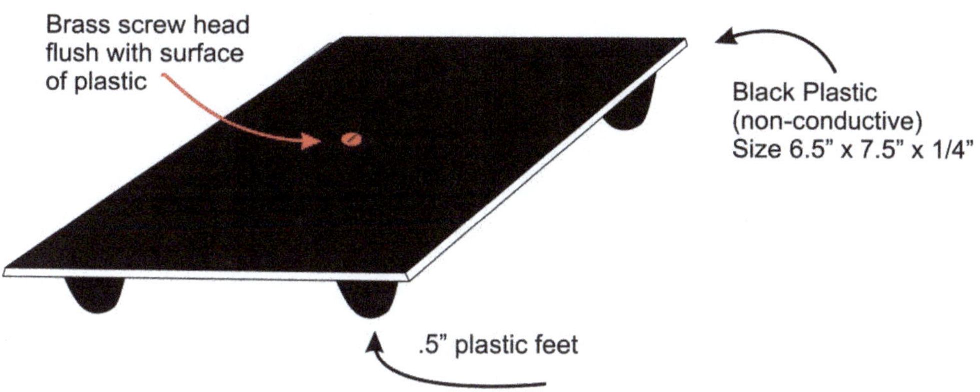

See Color Plate Online:
www.imagesco.com/Books/color-plate-pg1.html

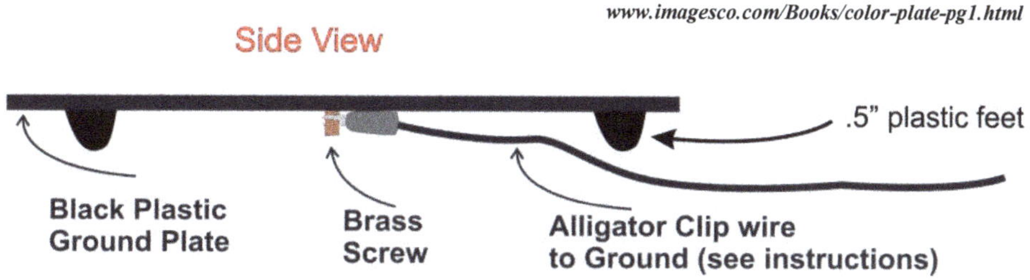

Figure 3-23 –Overview of Ground Plate

This ground plate is made from a piece of black (non-conductibve) plastic 6.5" x 7.5" x 1/4" thick, a brass pan-head machine screw with a brass washer and brass machine nut. For feet use four rubber adhesive feet that are a minimum of 0.5" tall. The size of the plastic sheet is not important, it should fit under your camera copy stand, see chapter 4. Let's start with a piece of plastic that is about 6.5" x 7.5". Although any size up to 8" x 10" will work. The black plastic is a background that provides a nice contrast to the corona discharge. The contrast improves the resulting Kirlian images.

It is important that the plastic be electrically non-conductive. Some black plastic materials use carbon (graphite) to obtain the black color of the material. Unfortunately carbon is electrically conductive. Having the ground plate made of this will short circuit your object and the corona discharge will not form around the object. This can be very frustrating and time consuming to figure out why you're not getting any image. I have years of experience shooting Kirlian photographs and it took me an hour or so to trouble shoot the problem back to the ground plate I was using.

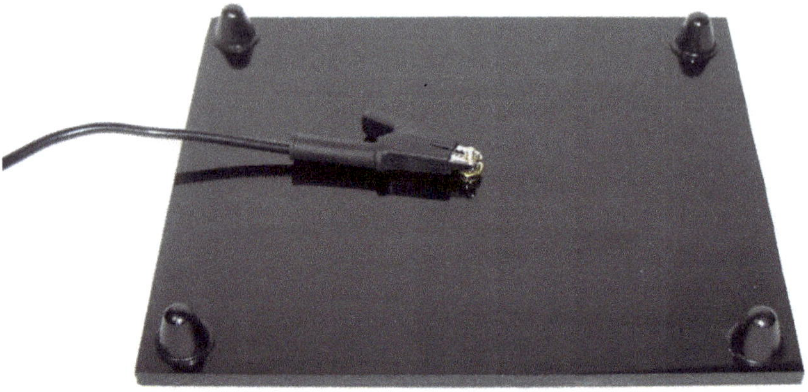

Figure 3-24 –Ground Plate Components

A 1/8" hole is drilled in the center of the black plastic sheet. This hole is counter sunk, so that the brass machine screw pan-head, lies flush with the surface of the plastic sheet. The machine screw is secured in the plastic sheet with the washer and machine nut.

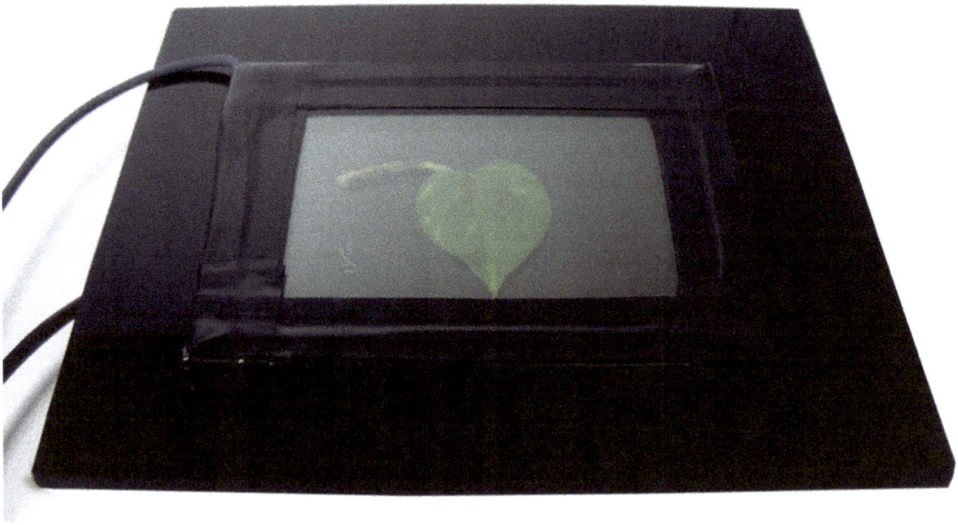

Figure 3-25 – Ground Plate, TDP and Subject, ready to make exposure

To complete the ground plate secure a piece of black non-conductive paper (black construction paper) or felt to the underside of the plastic sheet. Cut a slit in the paper to pass through the copper tape.

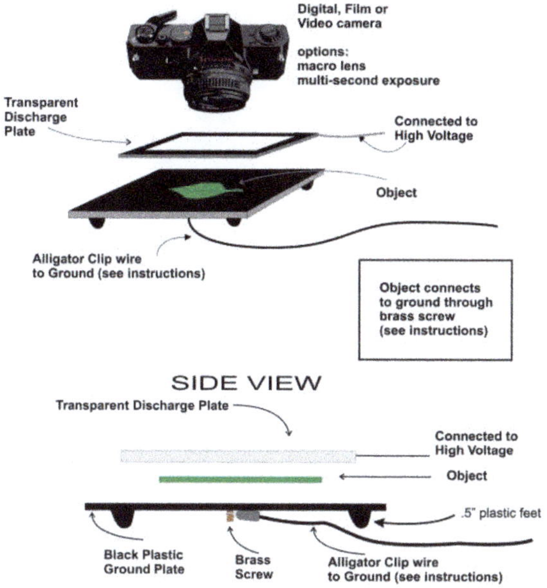

Figure 3-26 – Overview of Kirlian Photography set-up.

In use the object to be photographed is placed on top of the brass screw head. An alligator clip wire is connected to the bottom of the screw to circuit ground on Kirlian device or an earth ground. The Transparent Discharge Plate is placed on top of the Object and ground plate as show in figure 26. The camera is positioned above the set-up ready to take a long exposure picture.

Chapter 4

Digital Camera, iPhone, Video Cameras and Copy Stands

In this chapter we will look at the specifications of the various cameras you can use to record Kirlian photographs and video. In addition I will look at a number of copy stands for the cameras.

The Digital Camera:

There are four features you need in a digital camera.

1) Long exposure times. The most important feature is the ability for the camera to take long exposures of ten seconds in length or more. Expensive digital cameras will have a “Bulb” or “B” setting that keeps the shutter to remain open for as long as the shutter button is kept pressed. The long exposure time is necessary to allow the faint light from the electrical discharge to accumulate.

2) Macro Setting. The macro setting on a digital camera allows you to focus close to the object you are shooting. This is important when shooting small objects like coins and leafs.

3) Manual Focus. The manual focus setting allows you to manually focus on a subject, rather than having the camera auto-focus on the subject. Manual focus is important because you will need to manually focus the camera on the subject under normal lighting. Once the camera has been focused on the subject, the normal room lighting can be turned off. Turning off the lights is essential to capturing the faint light given off by the electrical (corona) discharge around the object we are shooting.

4) Prevent Built-In Flash. If your camera has a built in flash it is essential to be able to turn it off when making long exposures.

Digital Camera Copy Stands - Getting Steady to Shoot

Since we need to shoot long exposures to capture the corona discharge you need to consider how to keep your camera steady during the long exposure. If the camera moves during exposure you will produce a blurred picture.

The solution to this is a camera copy stand. I have a camera copy stand that allows me to have my camera facing down, so I can work with objects horizontally. I feel setting up the work on a horizontal plane provides the most stability. Especially when you're working in a dark room with a HV power supply, you also want your objects as stable as possible. The object should move during the exposure.

Inexpensive copy stands may be purchased for $10.00 to $25.00. A basic generic camera copy stand that is compatible with digital cameras, iPhone and Video cameras is shown in figure 1 below.

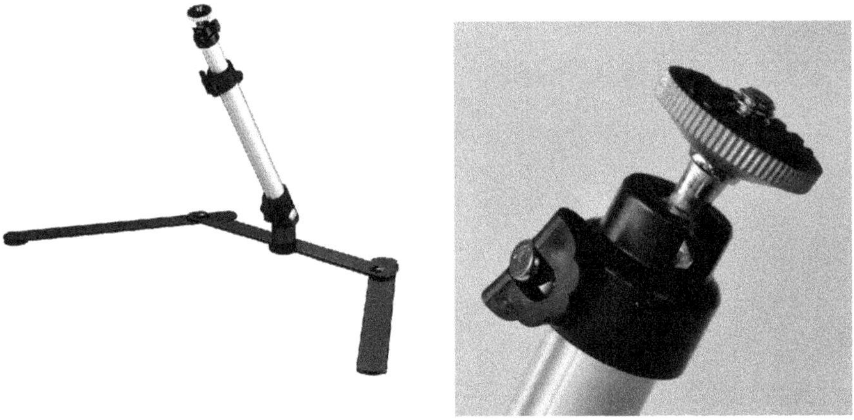

Figure 4-1 – Inexpensive Copy Stand Holder compatible with Digital Camera, iPhone & Video

The digital camera will have a mounting hole on the bottom where you can attach the camera to the top of the copy stand, see detail.

Tripods will work if you hold your discharge plate vertical to the camera plane. Some tripods allow you to put the camera holding shaft in from the bottom of the tripod, instead of the top. In doing so, you can adjust the camera so that it is facing to shoot straight down on the horizontal plane of the floor. Doing this allows you to set up you background, object and transparent discharge plate horizontally.

My standard digital camera copy stand is a large heavy duty unit that is very stable.

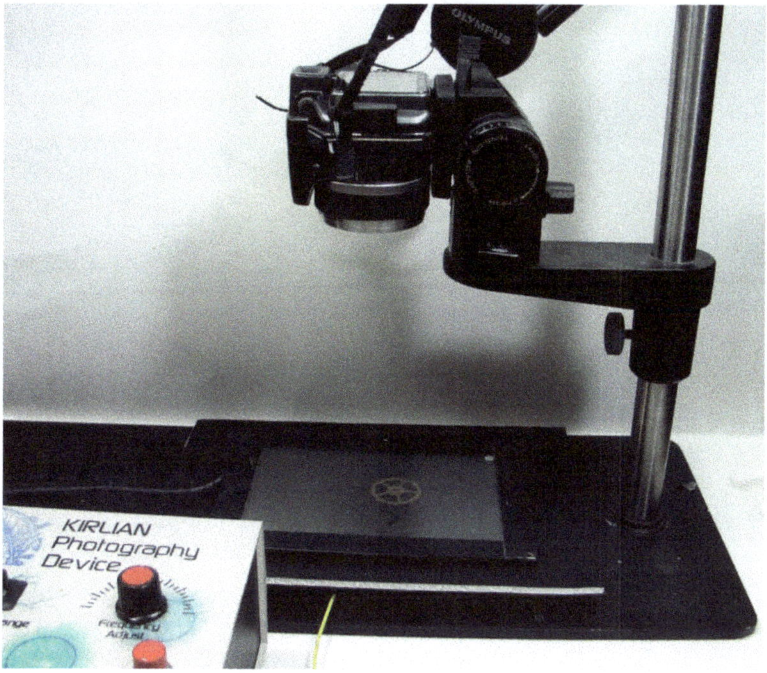

Figure 4-2 – Standard Heavy Duty Copy Stand

iPhone Camera:

I have the iPhone 5, so the information I am providing is valid for that particular iphone. However I am sure that if the application I use on my phone is available on your phone it will work also. To use the iPhone I use an app called "MagicShutter". This application allows you to lock focus on an object and manually set your shutter speed and therefore the camera's exposure time.

iPhone Camera Copy Stand - Getting Steady to Shoot

As with the digital camera we need to shoot long exposures to capture the corona discharge. So you need to consider how to keep your iPhone camera steady during the long exposure.

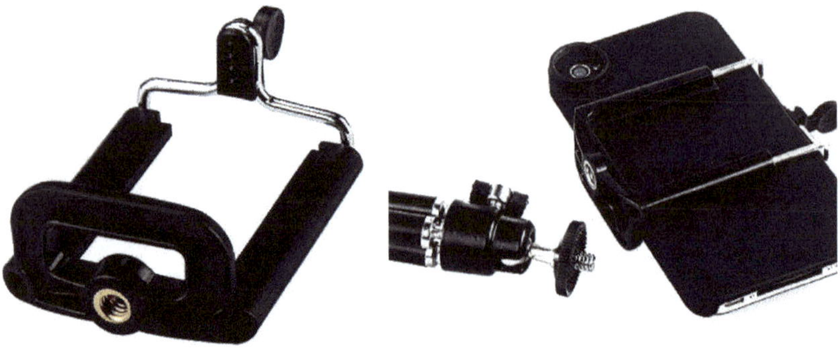

Figure 4-3 – iPhone adapter to secure iPhone to Copy Stand.

There are two solutions, we can use our digital camera copy stand with an adapter to hold the iPhone . The iPhone adapter allows you to clip your iPhone onto the adapter which can screw into a copy stand screw connector.

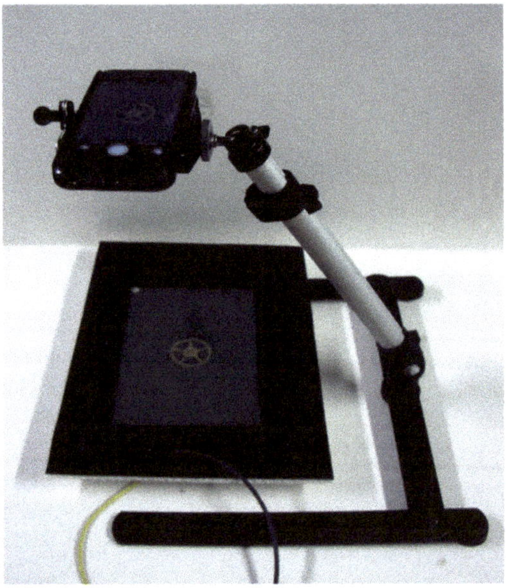

Figure 4-4 – iPhone, adapter, Copy Stand Overview.

Making a Cardboard Box Copy Stand

Copy stands don't have to be complicated devices. You can make a useable copy stand out of a small cardboard box. To illustrate this, I am providing directions to make a copy stand box for the iPhone. You can use a similar process to make a copy stand for your particular camera.

I begin with a 7" x 9" x 3" white box.

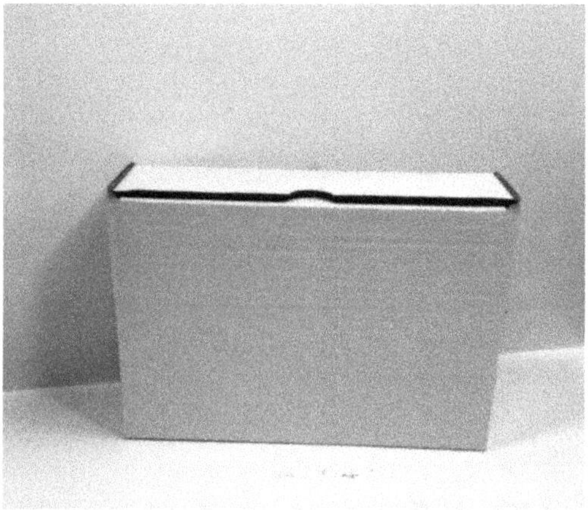

Figure 4-5 – 7" x 9" x 3" White Cardboard Box

The 3" width is important; it is wide enough for my iPhone to rest on. I taped the edges of the box, so it will not open up in use and to make the box more sturdy.

Next I mark the two sides of the box with a pencil, see Figure 4-6.

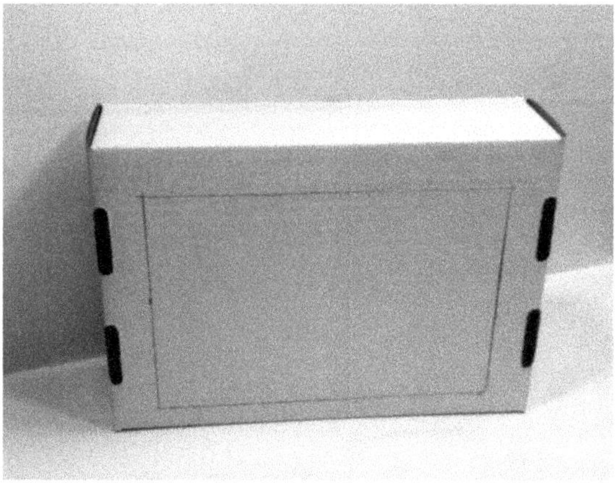

Figure 4-6 – Marking the sides to cut out

I leave a 1" border from the sides and top, and ½" board on the bottom. The bottom of the box, is the side where the top of the box folds into.

Next I cut out the panels on each side of the box.

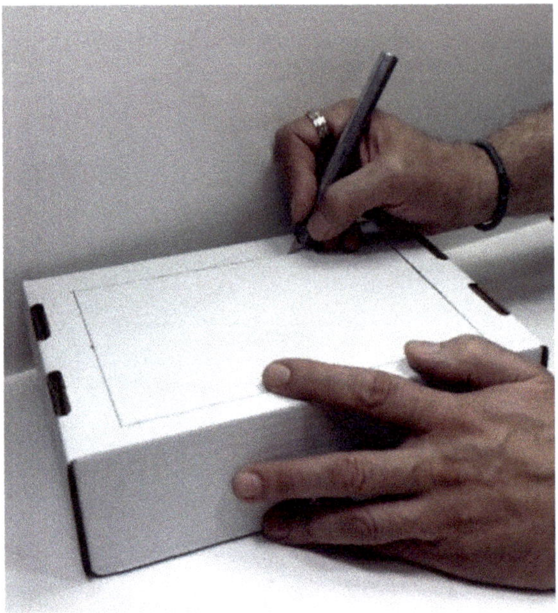

Figure 4-7 – Cutting the panels out with an exacto knife

With the panels cut I trace the outline of my iPhone on the top of the box. I mark the area where the camera lens is position inside the outline and I cut an opening to allow the camera to look down at the inside bottom of the box.

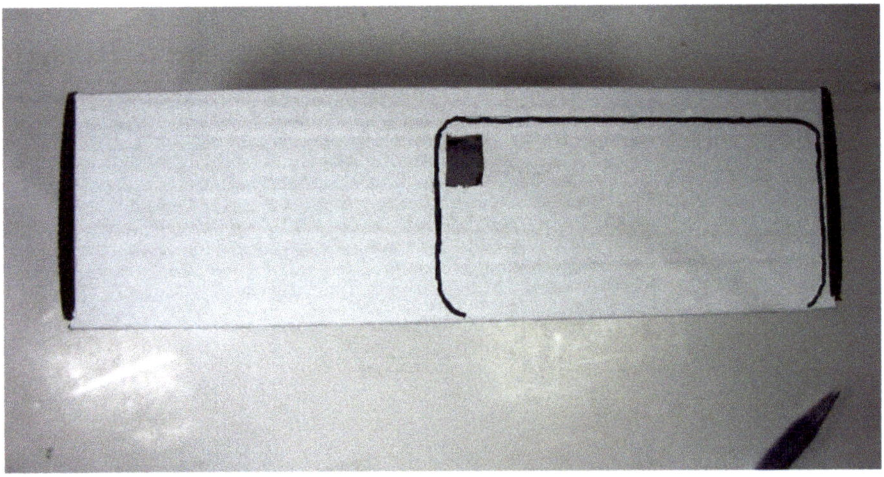

Figure 4-8 – Marking the iPhone outline and cut out for camera lens

The ground plate and the transparent discharge plate can be placed at the bottom of the box. The iPhone is placed on top of the box ready to take a picture.

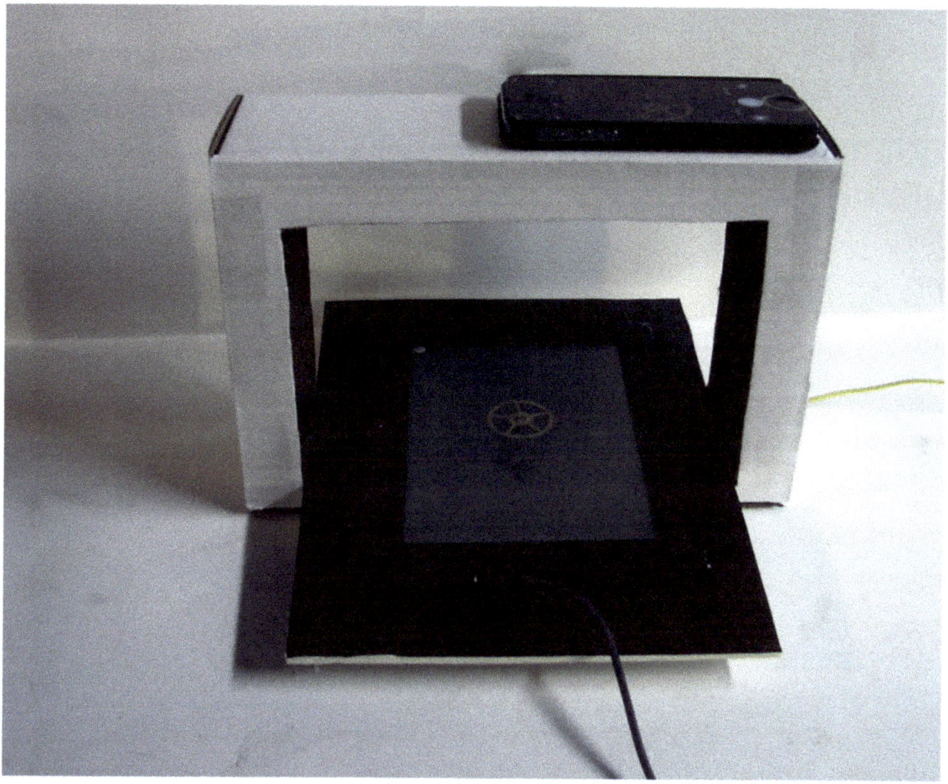

Figure 4-9 – Cardboard iPhone Copy Stand

iPhone Apps

In order to use the iPhone to take Kirlian photographs you need to purchase an app that will allow you to take long exposures. From what I read online, these apps don't actually keep the iPhone shutter open, but instead take multiple pictures and layer the pictures on top of one another accumulating the light gathered in each individual photograph into the finished picture.

This is a similar technique used in astronomical photography.

The app that I am using is called "Magic Shutter". There are two others apps "Slow Shutter Cam" and "Long Expo". I have not tried these other apps but reading their specification they seem to me that they will work.

Video Camera

There are a number of video cameras on the market capable of shooting in the low light generated by the corona discharge. These are sold as low lux video cameras. You need a true low lux, you do not want a camera that enters a slow or still frame mode to gather light. You do not want an infrared night vision camera.

The two video cameras below I have personally used and have obtained good results with.

B/W Video Camera

Low lux B/W (black and white) video cameras are less expensive than color. You want a camera that can see in 0.0003 lux like the KT&C Sony CCD Model# kpc-ex190-swx. This camera will cost around $100.00 (2016 price) including a lens.

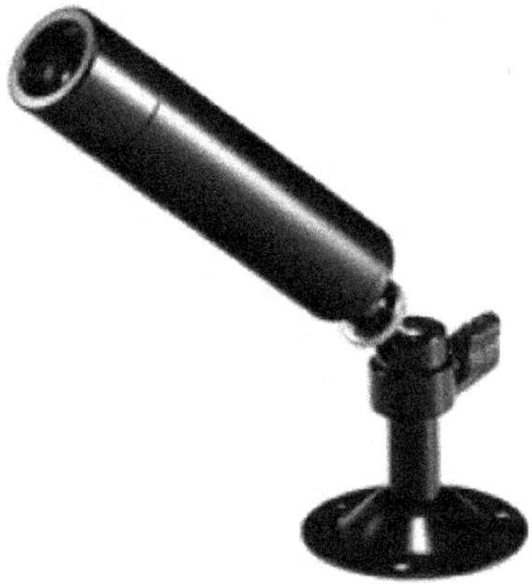

Figure 4-10 – B/W Video Camera

Sony 1/3" ExView B/W CCD
420TVL Horizontal Resolution
Automatic Gain Control (AGC)
Water-Proof (IP67 Rating)
Automatic Backlight Compensation (BLC)
Low Illumination of 0.0003 Lux / FacePlate
Built-in Reverse Power Input Safety Circuit
Ultra Compact Size of 0.75"(™) Dimension

Color Video Camera

The Sony KPC-650 Color is a color video camera that can see in 0.05 lux. Not as sensitive as the B/W but good enough to work. This camera costs about $200.00 (2016 price) without a lens. Add about $50.00 more for a lens. When choosing a lens I recommend getting a lens that will allow the camera to focus on a 5" x 5" area at a close distance of 12" or less. This will allow the camera to be close to the TDP and obtain good detail of the object you are photographing.

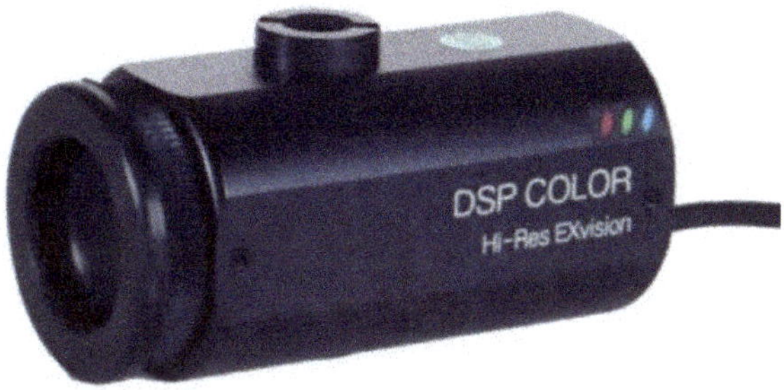

Figure 4-11 –Color Video Camera

Color Box Type Camera
Sony 1/3" ExView Color CCD
550TVL Horizontal Resolution
C/CS Mount / Vari-Focal C/CS Mount Lens
Min. Illumination of 0.05 Lux / Faceplate
Automatic Tracing White Balance (ATW)
Automatic Gain Control (AGC)
Backlight Compensation (BLC)

Newer Video Cameras

The video cameras I have listed are older models that I have personally used. An internet (or ebay) search for low lux cameras will turn up newer cameras that are more light sensitive and less expensive than the cameras I have listed. I have seen color cameras like KPC-650BH working at 0.0003 lux and B/W cameras like the KT&C 750TVL working down to .000001 lux. I am sure these video cameras are fine, but without having personally used these cameras I am reluctant to recommend them.

Chapter 5

Shooting Kirlian Images

No Lights, Action, Camera

What was that? No lights? Yes that's right, no lights. In order to capture and record the feeble light of the corona discharge, namely Kirlian images, you need close to complete darkness when making your exposure. How do we do that? We have a few options, shooting in a light tight room or using a light tight cover over your camera, subject and transparent discharge plate.

The simplest option is to shoot in a windowless room, where you can turn off the lights and have relative darkness. If the room you are planning on working in has a window, you could black out all light from the window or maybe you could shoot Kirlian photographs at night. Working at night makes it easier to block any (artificial) light coming in from a window. The big disadvantage is you have to work at night.

The second option is to build a light tight cover for your Kirlian equipment so you can shoot in a lighted room. And there are a few ways to do this also. The simplest is simply to set up the shot and to drape a light tight black cloth over the equipment and shoot. Near the end of this chapter I will show you how to build a simple light tight enclosure for less than $10.00 out of black foam board that will cover most equipment.

The following picture taking information is assuming you're working in a light tight (windowless) room, or a windowed room at night. Please note the procedures for shooting images is almost the same when using a light tight cover. The differences will be cover later.

Shooting using Digital Camera

Set up your camera, copy stand, ground plate, object, transparent discharge plate and high voltage power supply, see Figure 5-1.

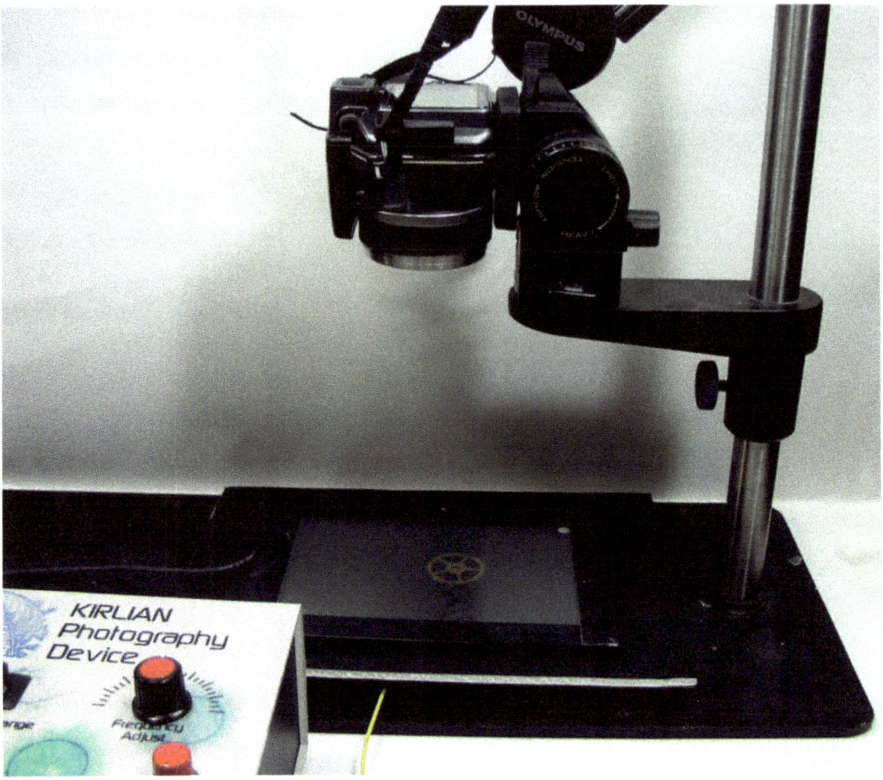

Figure 5-1 – Digital camera and copy stand set up.

Attach the high voltage lead from your circuit to the transparent electrode. If you are photographing an inanimate object such as a leaf or coin, attach a ground wire to the object. A black background usually works best for Kirlian photographs, so place black (non-conductive) paper on the table where you will be shooting. Then place the object on top of the black paper, and finally the transparent discharge plate on top of the object.

When placing the transparent discharge plate on the object, place the plastic sheet side toward the object. This way the corona discharge only needs to travel through the .010 (.254 mm) thickness of plastic. If you place the discharge plate glass side toward the object, the corona discharge would need to travel through .125" (3.2 mm) of glass, making the brightness of the corona discharge much fainter.

If you purchased a Transparent Discharge Plate, the thin glass side has a colored dot on it. Place that side toward the object.

Situate your camera above the object and frame the object as you want it to appear in the finished photograph. Set you camera to manual focus.

Then manually focus on the object with the room lights on. If you have the option of setting your F-Stop, open your aperture as much as possible. You're shooting a flat object, you don't need too much depth of field. Next set the shutter speed, set it to an long exposure time say fifteen seconds to start is ideal.

Turn off the room lights. Turn on the power to your high voltage circuit. If your HV power supply has a frequency adjustment now is the time to adjust the frequency to provide the best corona discharge. When you have the best corona discharge, then hit the camera's shutter. Keep high voltage power applied for the full time of the exposure (15 seconds in this example). Sometimes you may want to vary the frequency during the exposure to record different corona effects the change in frequency may cause.

After the exposure you can turn on the room lights and evaluate your Kirlian photograph. You can, adjust (exposure time), for instance if the corona is too bright and blurry, you can reduce the exposure. Or if the corona wasn't bright enough you can increase the exposure time   and re-shoot.

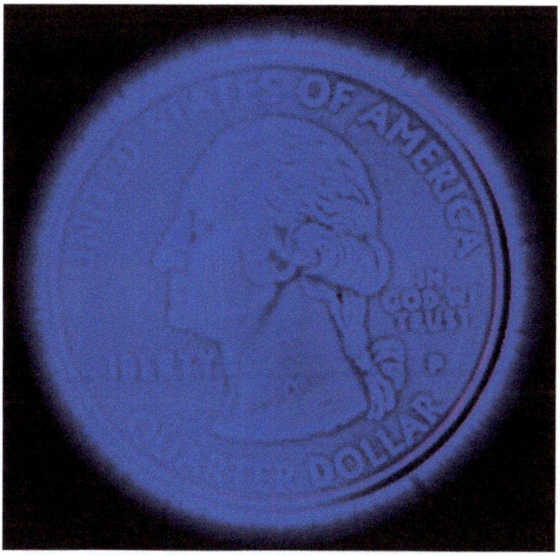

Figure 5-2 – Digital camera picture.

In general, because we are shooting in a nitrogen rich atmosphere, the corona discharge is a blue-ultraviolet color, and is recorded as such onto our equipment. The pretty colors, vibrant reds, oranges and greens obtain with sheet film is false coloration caused by the activation of the color dye layers as the corona discharge travels through the film's dye layers. While digital photographs are not as colorful as sheet film renditions, the corona discharge captured through the transparent discharge plate is more accurate. In addition, the photograph can be imported into software such as Adobe Photoshop and recolor the corona discharge.

Shooting using iPhone

The procedure is pretty much the same as with the digital camera. Use whichever long exposure app you purchased to set the exposure time. With the room lights on allow the camera to focus on the object/subject. If you purchased the Magic Shutter App you also have the ability to lock the focus on an object. Lock the focus. This will insure your object stays in focus when you turn off the room lights. This is a great feature, almost as good as manual focus.

Adjust and frame the image of the object so that if fills the camera's screen as you see fit. One disadvantage of the iPhone's Magic Shutter App is that you cannot expand an object in the view to allow that portion of the image to fill the screen. But that's okay. When you turn off the room lights the camera will keep the object framed and focused as you set it.

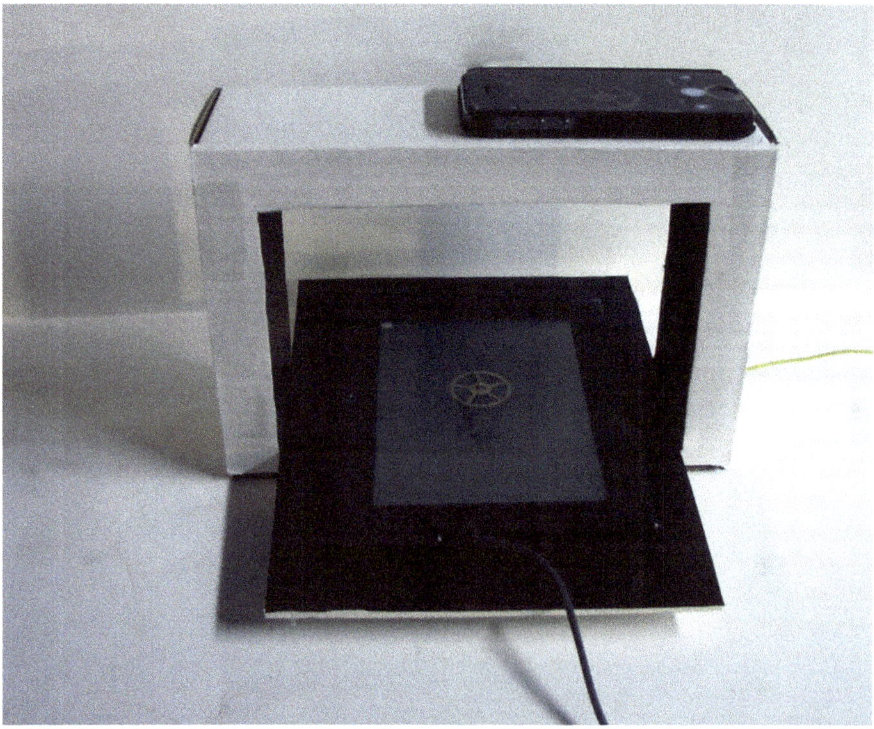

Figure 5-3 – iPhone camera and copy stand set up.

Turn on the power to your high voltage circuit. If your HV power supply has a frequency adjustment now is the time to adjust the frequency to provide the best corona discharge. When you have the best corona discharge, then hit the iPhone's shutter. Keep high voltage power applied for the full time of the exposure (15 seconds in this example). Sometimes you may want to vary the frequency during the exposure to record different corona effects the change in frequency may cause.

After the exposure you can turn on the room lights and evaluate your Kirlian photograph. You can, adjust (exposure time), for instance if the corona is too bright and blurry, you can reduce the exposure. Or if the corona wasn't bright enough you can increase the exposure time and re-shoot.

Figure 5-4 – iPhone camera images.

The main reason the iPhone picture of the quarter (Figure 4) is not as high resolution as from the digital camera is that I could not zoom in on the quarter. The image in figure 4 is cropped from a full view image is shown in Figure 5-5 below.

Figure 5-5 – iPhone camera full picture view uncropped.

There are two obvious solutions to this problem. One is too position the camera as close as possible to the subject (quarter) as long as the iPhone's camera can maintain its focus. Or one can purchase a macro lens adapter for the iPhone camera.

Shooting Real Time Video

Shooting real time video through the transparent discharge plate is also possible. However it requires the use of a extremely low lux video camera or a standard video camera retro-fitted with a light amplification unit as discussed in chapter 4.

We can use the same set-up for shooting video as was used for the iPhone and digital camera. The camera is mounted above the TDP set-up. This will provide perfectly good real time video.

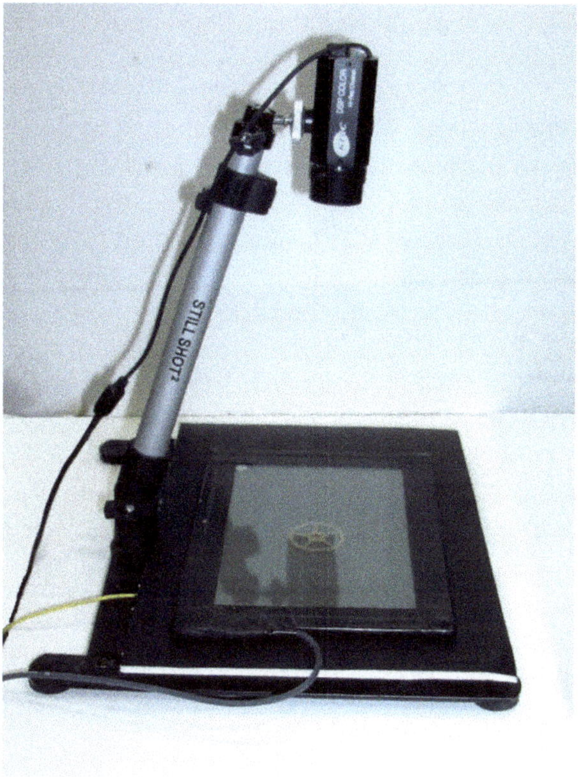

Figure 5-6 – Color video camera set-up.

For anyone interested in pursuing this type of photography you can start by looking at this Youtube tutorial video I created:

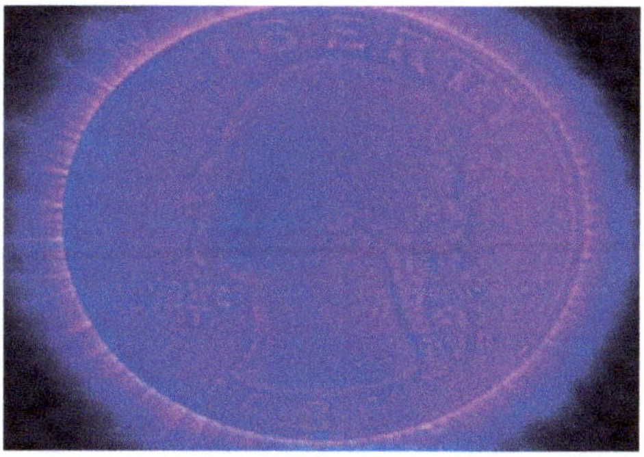

Figure 5-7 – Still frame from video camera.

https://www.youtube.com/watch?v=FLU3agfVGfE

Shooting People's Kirlian Aura's

Sooner or later you are going to want to make kirlian photographs of people. Do not make a Kirlian exposure of anyone with a heart condition or any type of implanted electrical device such as a pacemaker.

To shoot a person's (or live subject) Kirlian aura start with the person's fingertip. Make sure the person you are photographing does not touch or come in contact with a ground while making the high voltage exposure. Grounding will cause excessive electrical power to flow through the person and cause an electrically shocked. Also do not place too much surface area of the person you are shooting on the transparent discharge plate (or film). For instance, placing the whole hand on the transparent discharge plate, will dissipate the corona discharge so much, that it is not practical to obtain expose the film even with a long exposure time.

When I shoot people's finger tips I reverse the position of the camera and transparent discharge plate. This allows you to capture the full corona discharge from the fingertip easily. If you were still shooting top down, the finger would block part or all of the corona discharge.

The camera is at the base pointing up toward the transparent discharge plate. The transparent discharge plate plastic sheet side is facing up, where one would be touching the plate. This way the corona discharge only needs to travel through the .010 (.254 mm) thickness of plastic. The image below shows the set I use to record real time video of fingertips.

Video Set-up

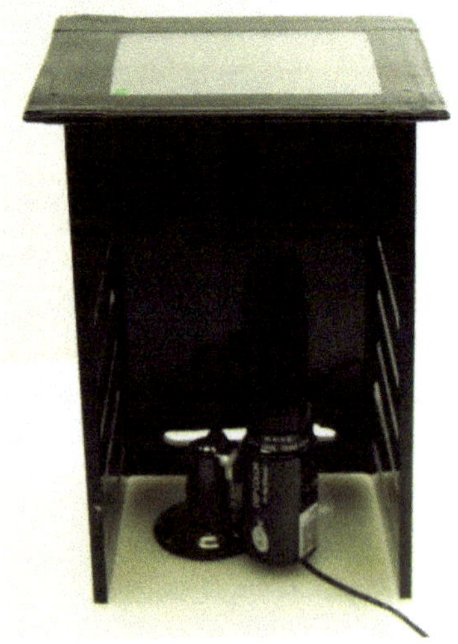

Figure 5-8 – Video camera set-up for shooting finger tips.

Figure 5-9 – Fingertip still frames from video camera.

Light Tight Covers

Using light tight covers presents a few challenges. First and foremost is triggering the digital camera or iPhone camera. Video is not much of a problem and the video can operate continuously. With digital cameras and iPhone you have two possible solutions.

Self Timer

First solution, set the self timer on the camera to about 60 seconds to make a timed exposure. This means that the camera will take the picture 60 seconds after you hit the camera shutter. If you are using the Magic Shutter App for the iPhone that I recommend, it includes a self timer that you can set. For your digital camera you need to look up the self timer in the manual. Most modern digital cameras has this feature.

Eternal Shutter Release (Cable)

Second solution is to use an eternal shutter release cable, or a more modern wireless shutter release. For your particular digital camera you need to check to see what eternal shutter release accessory is available. For the iPhone you can use a headset with volume controls as an eternal shutter release cable.

When the iPhone is in camera mode, hitting the volume control on the headset, triggers the shutter. You can purchase an inexpensive headset for the iPhone, and clip off the ear buds to make a dedicated shutter release cable. A nice Bluetooth wireless option is the MuKu Shuttr,

http://www.amazon.com/gp/product/B00JK5OW8C/

Simple Soft Light Tight Cover

A simple soft light tight cover is black light tight material that can be draped over your equipment to cover it. Black velvet material is perfect for this. You set up your shot, set manual focus and exposure. Cover the camera and transparent discharge plate with the black material, then turn on the HV power supply and trip the camera's shutter. When the exposure has been made, turn off the HV power supply, remove the cover and check the resulting image.

To improve on this design and if you can sew, cut the black velvet into four or five shapes that when sewed together form a simple tent. Then sew the pieces together.

Simple Light Tight Cover from Black Foam Board

A simple light tight cover can be made from black foam board. Black foam board come in a number of pre-cut rectangles. These foam boards are extremely light and durable. I hot glued the pieces together. The method that I found to be the best way to hot glue the foam board is to run a bead of hot glue along the entire edge of one piece of foam board. See the following sequence of drawings. Start with five pieces of black foam board, the size I used was 11" x 14" x 3/8" thick. Glue two boards together as shown below.

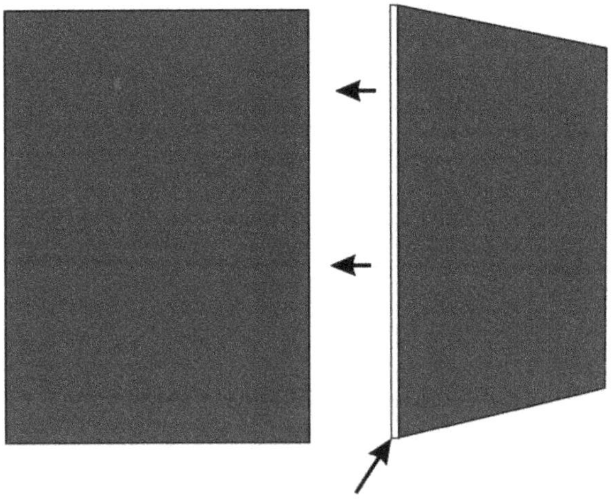

Figure 5-10 -- Construction of light tight cover (1)

Place a bead of hot glue along the edge of one side of the black foam board. Then join that piece with another piece as shown.

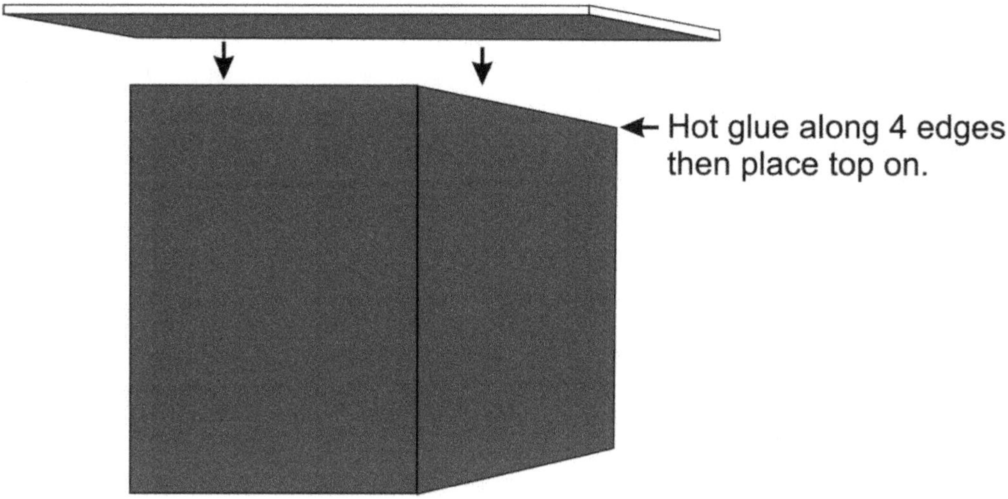

Figure 5-11 -- Construction of light tight cover (2)

Continue in this manner until you have four pieces joined together, creating an open ended box. Then place a bead of hot glue along the top edge of the four sides, then place the top down securing it to the four sides.

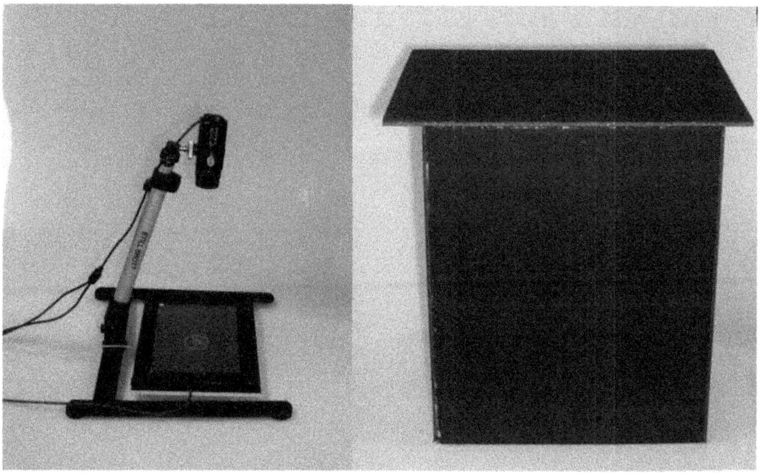

Figure 5-12

The light tight box fits over the camera setup and transparent discharge plate.

Making the Light Tight Box Light Tight

If the hot glued edges of the light tight box, leak light, cover the outside edges of the box with black electrical tape.

When using the light tight box, depending upon your set-up, you may have wires or cables coming out from underneath the box which may cause some light to leak in from that side. I recommend cutting a small notch in the bottom of one side of the box. The notch should only be large enough to allow the wires and cable to pass underneath without lifting the box. The get a piece of black clothe or black velvet and securing a few inches of the material near the bottom of the light tight box so that it drapes over the notch in the bottom of the box and extends a few inches out from the box. This will seal any light from leaking in from any wires or cables coming out.

Light Tight Box - Try Out

I set up my iPhone camera to reshoot the quarter Kirlian image. I used the light tight box and an eternal shutter release to the iPhone. The Magic Shutter iPhone app. was set to 8 seconds. I moved the quarter as close to the camera lens as possible. Here are my results:

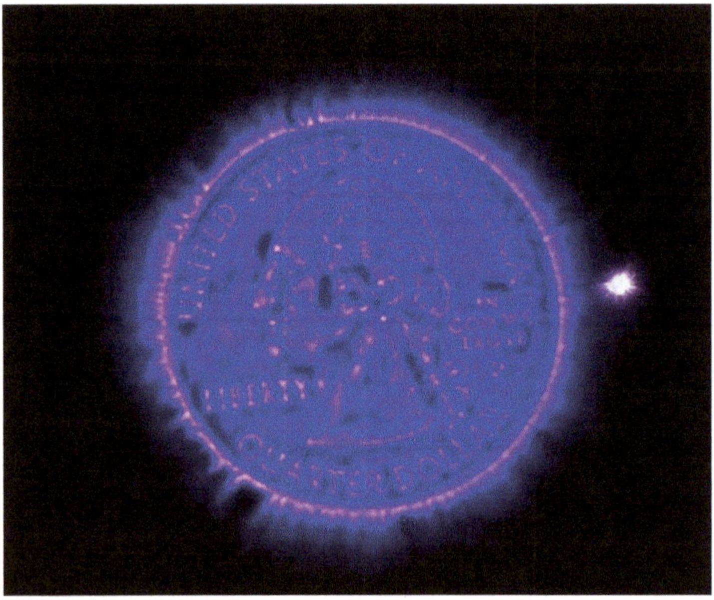

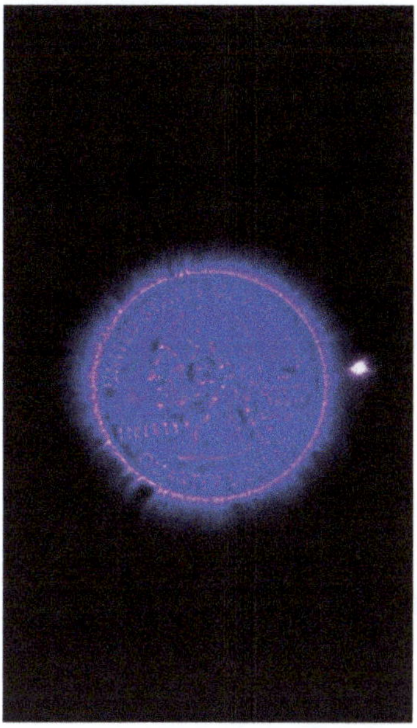

Figure 5-13

The iPhone Kirlian image is a significant improvement over the first iPhone Kirlian image. The image on the left is cropped and the image on the right is un-cropped. You can compare the size of the subject in these images to the size of the subject in original images in figure to see how much closer I was able to photograph the quarter. I did forget to clean the iPhone camera lens. This image can be improved further by using a macro lens for the iPhone.

90' Mirror Set-Up

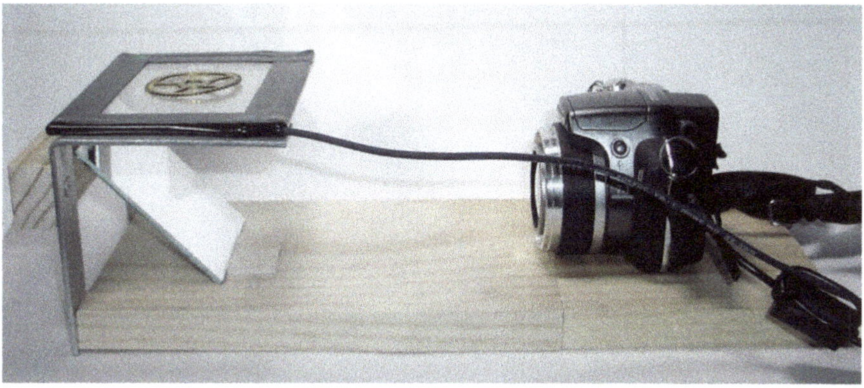

I created a 90 degree mirror set-up device that is useful in some Kirlian photography situations. I use this mirror set-up in a dark room. It's not always necessary to have complete blackness to take Kirlian images. I have take finger print images as well as a host of other objects.

Same grounding procedures are used. Inanimate objects like coins should be ground to an earth or circuit ground. Animate objects like people when shooting their fingertip, should not be grounded or touch a ground when making a kirlian exposure.

The mirror used in this is a front surface mirror. A front surface as its name suggests has it reflection coating on the front surface of the glass. This is in contrast to a standard mirror whose reflection coating is on the back surface of the glass. Having the reflection coating on the back surface allows the glass itself to protect the reflecting surface. While this is good for most applications, like looking at yourself in a mirror, optically it is not ideal. As the front surface of the glass will also reflect a faint image.

As a try-out you can use a standard mirror, but you may see some ghosting of a faint image reflected by the surface of the glass in your images.

The front surface mirror eliminates that double reflection.

The components are shown in the following photograph.

(2) 4" Steel Corner Brackets

(4) Wood Screws

(1) Front surface mirror 3" x 3.5"

(1) Pine wood ¾" x 5.5" x 12" Base

(1) wood piece ¾" x 5.5" x 1.25" Back Stop

Misc. small piece of plastic 1/8" thick x 1" x

1.5”.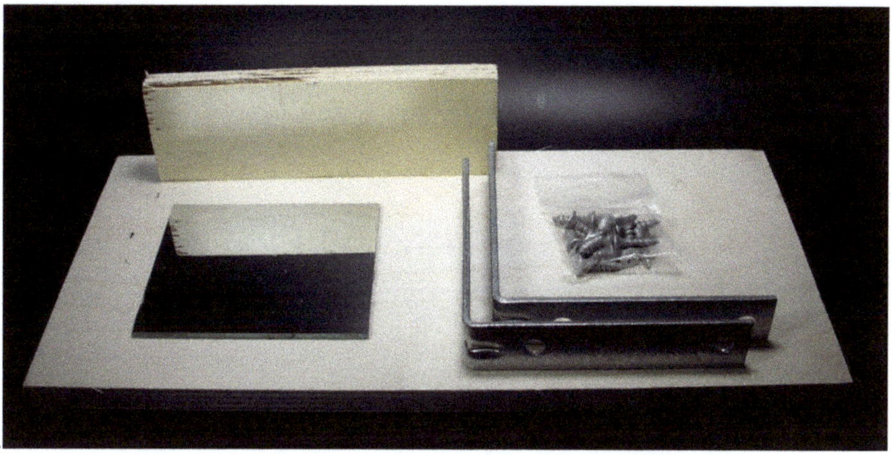

The 4” corner brackets are screwed into the 5.5” x 12” base.

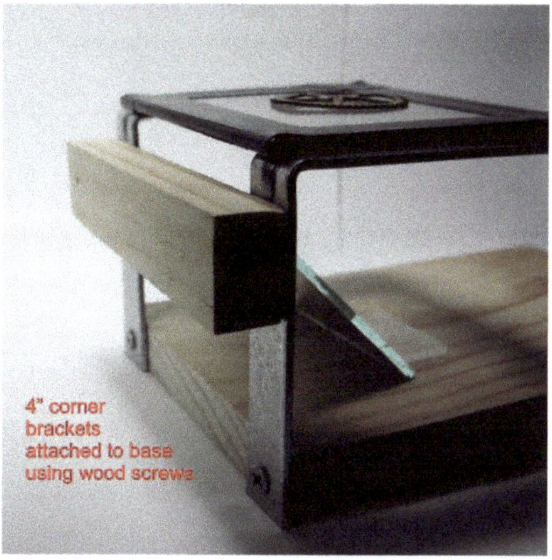

Next the back stop wood is attached to the 4” corner brackets. The plastic stop is hot glued in a position that will put the front surface mirror in a 45 degree angle. The front surface mirror leans on the back stop wood, with its front edge resting on the plastic back stop. The front surface mirror is not lock into place, it can be easily removed for storage or cleaning.

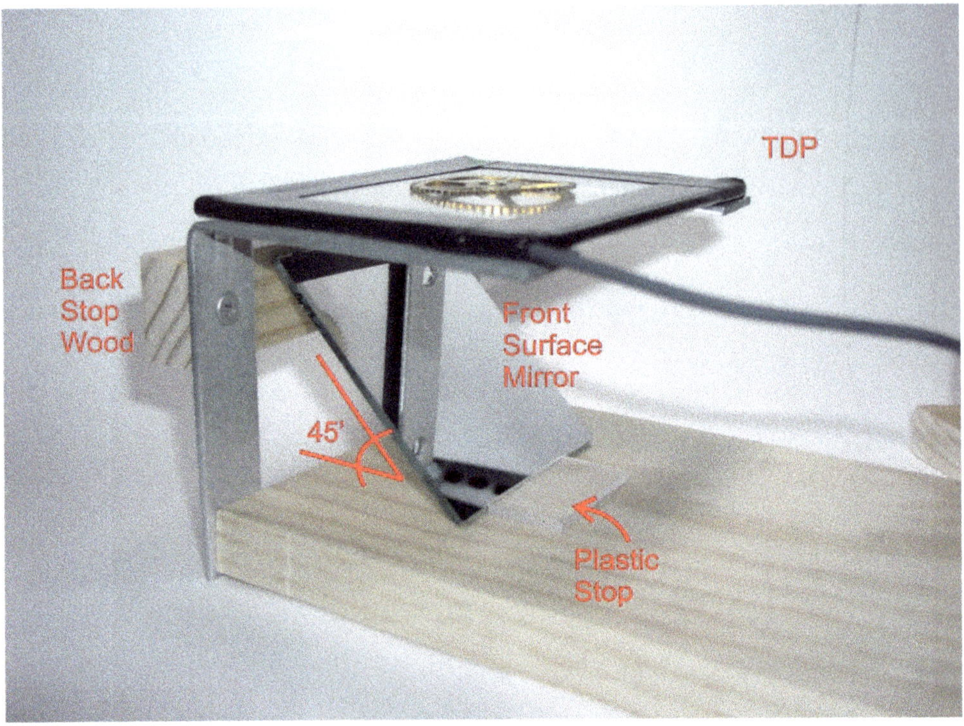

Positioning your digital camera in front of the front surface mirror, and using a macro setting on your camera, will allow you to capture the image of an object resting on the TDP in great detail.

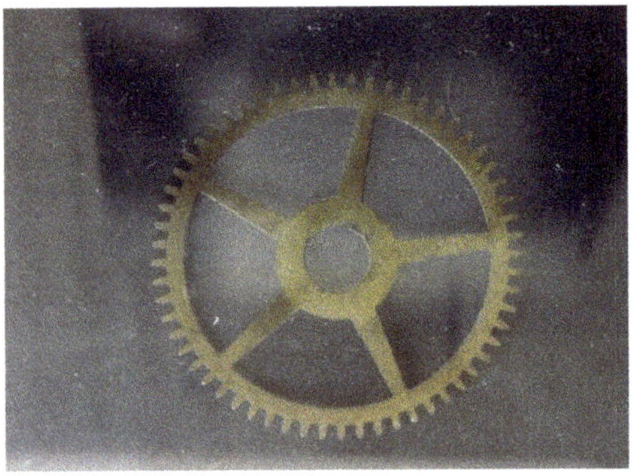

What to Shoot

Now that you know how to shoot Kirlian photographs the question evolves into what do I shoot. Most small metallic objects are good subjects, coins are an excellent example, but paper clips, gears, earrings etc work well. Plants and leaves are favorites of Kirlian photographers, as are wedges of fruits and vegetables. Typically you are looking for subjects that have some electrical conductivity; whether that conductivity comes from metal (coins) or water (leafs) doesn't matter.

There is also a list of what not to shoot, or the difficult to shoot. Nonconductive objects are difficult to shoot, these include subjects like plastic, rocks and hair.

Chapter 6

Clearing up Noise in Digital Photographs:

My digital camera used for the majority of this book has a maximum shutter exposure of 16 seconds. In this exposure time the photograph captured on the camera's CCD or CMOS element accumulates a little digital noise. This digital noise consists of random pixels of various colors; gray, reds, and yellows sprinkled throughout the resulting photograph. You can remove a large amount of the digital noise if you have access to a photo-retouching software like Adobe Photoshop 5.0 (or later) or a similar photo-retouching product.

Here are the steps I used to remove noise:

1) Starting with base image of leaf: Select Image → Adjust → Brightness/Contrast (See figure 6-1.)

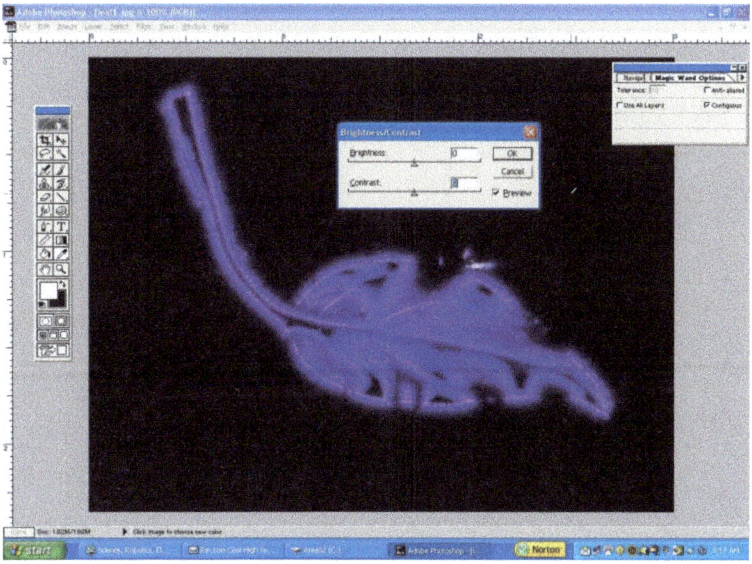

Figure 6-1

2) Increase contrast to brightened image, see Figure 6-2.

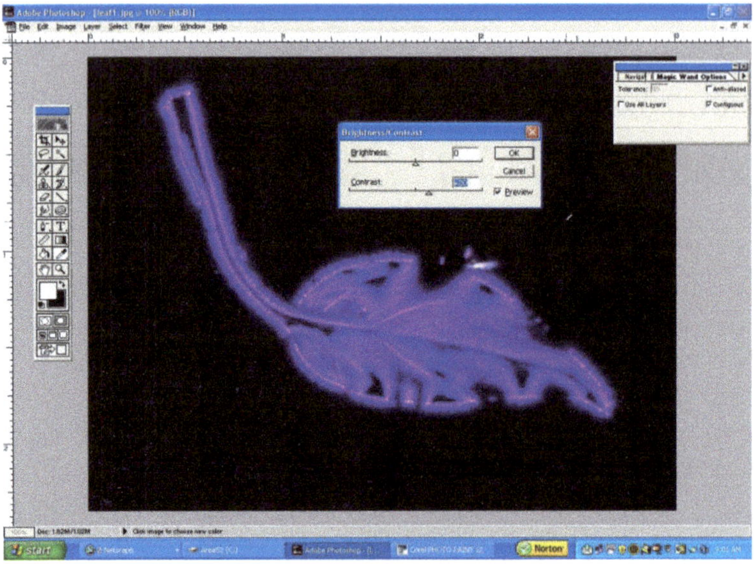

Figure 6-2

3) Zoom picture to 500% so you can clearly see the background pixel noise. Using the Magic Wand (set want to 30 tolerance), click on a noise pixel. In figure 6-3. I choose a red noise pixel.

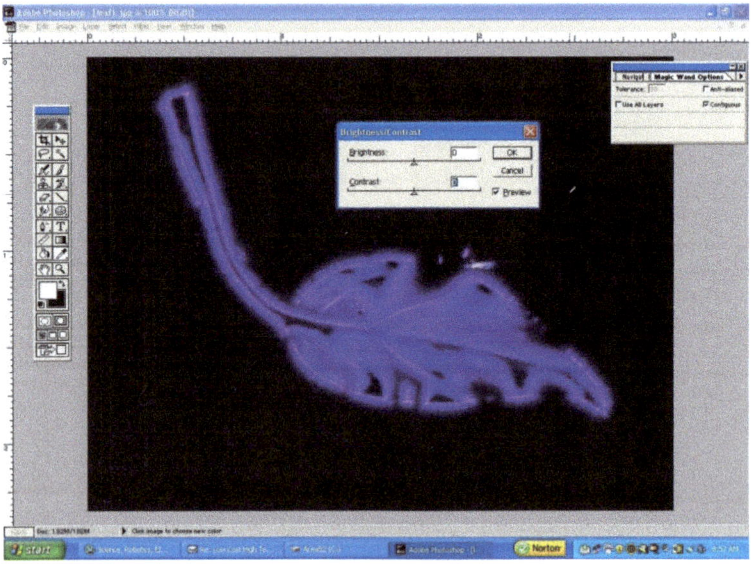

Figure 6-3

4) With the noise pixel selected, go to Select → Similar. Now all the pixels with that color in the photograph will be selected. See Figure 6-4

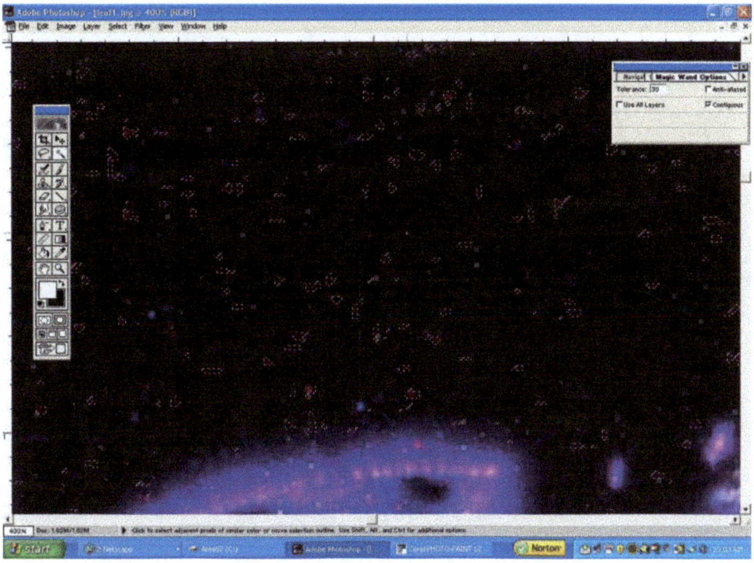

Figure 6-4

5) Set foreground to black and hit the delete key. All the selected noise pixels will be colored black. See Figure 6-5.

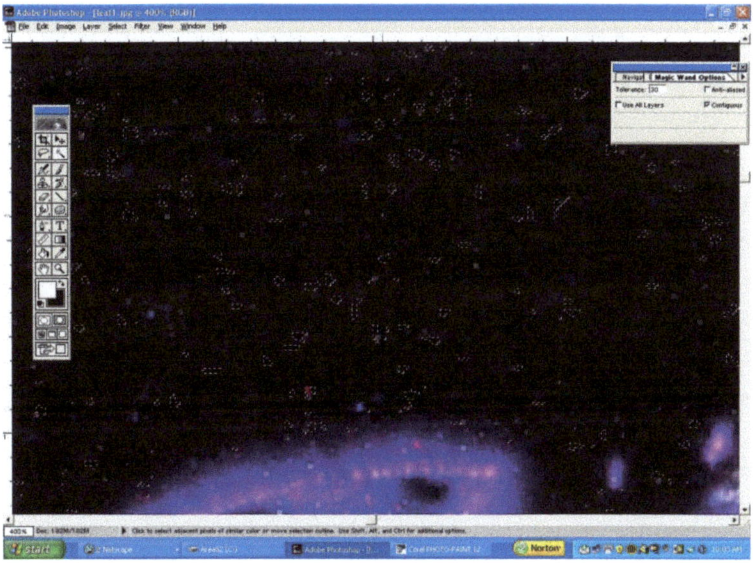

Figure 6-5

Continue in this manner, clicking on various colored noise pixels, Selecting similar and deleting them to black. Be careful not to select any noise pixel that is the same color or shade of color used in the main image you are trying to enhance.

Figure 6-6 shows a leaf section with the noise background in the original photograph and cleaned up.

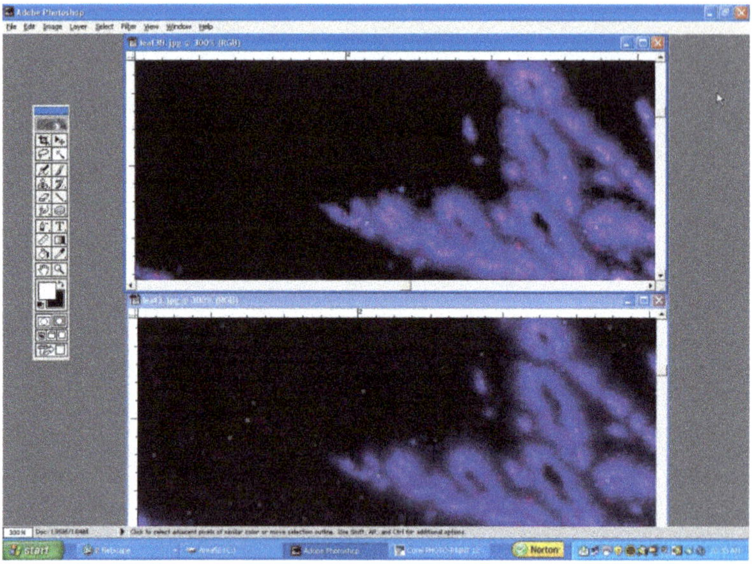

Figure 6-6

Dark Field Noise Removal:

A second method to remove noise pixels is to first create a "Dark Field" image. You will need Photoshop or an image processing software like Photoshop to use the Dark Field.

It has been my experience that the longer the Digital camera is left on, the more hot pixels show up in the dark field. Literally as the CCD element becomes warmer, the more noise pixels there are.

Dark Field Images

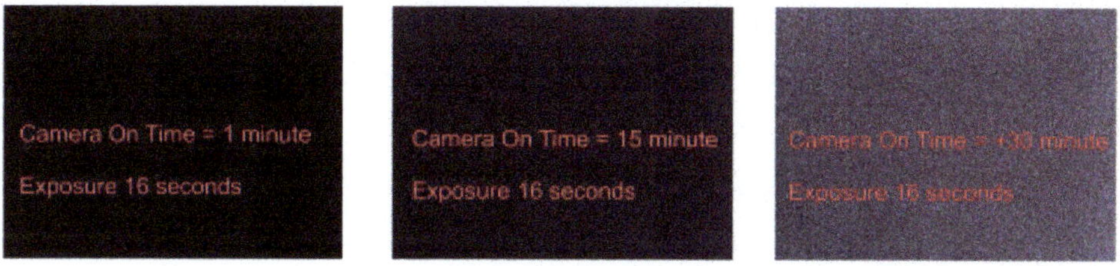

Camera On Times are approximate:
All three images brightness was increased to visualize hot pixels

To create a dark field image, keep the digital camera's lens cover on when you make a long exposure. If you can't keep a lens cover on, when the camera is on, then cover the lens with a black cloth. Then set the camera to a long exposure time, and make an exposure. The resulting image will have color pixels sprinkled throughout the image. This is the camera's hot pixel map.

Save this image, name it Dark-Field.

Now shoot long exposure Kirlian images.

Open your original image and your dark frame in Photoshop.

Select the "Move" tool from the tools palette (or press "v").

Click and drag the dark frame over into your original image while holding down the Shift key. By holding Shift, the dark frame will center itself automatically on top of your original image. This is important so that the hot pixels will be aligned perfectly in both images.

On the layers palette, change the blend mode of the dark frame layer to "Difference". At this point, you should no longer see the hot pixels in your image. You can adjust the Opacity of the dark frame layer until the image looks good to you.

Pixel Zap Software:

Pixel zap is a software solution to noise pixel removal. This software can use either an algorithm or dark-field image for hot pixel removal. The software is available here:

http://tawbaware.com/pixelzap.htm

The free shareware version of pixel zap is limited to a 640 x 480 pixel image. If you purchase this software the image size limitation is removed.

Coloration:

Your photographs are mostly blue-violet with a little white. This is due to the fact that the corona discharge in nitrogen (air is 78% nitrogen) generates blue to UV light.

You can use the color balance feature of Photoshop to change the color of your image.

Select; Image → Adjust → Color Balance.

Change the slider for the colors you can adjust the color balance for the; Shadows, Midtones and Highlights.

Figure 6-7 is the starting image.

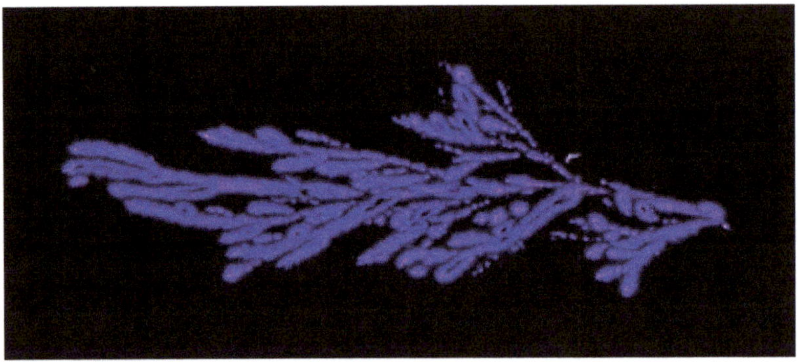

Figure 6-7

Figures 6-8 and 6-9 have their color balance manipulated.

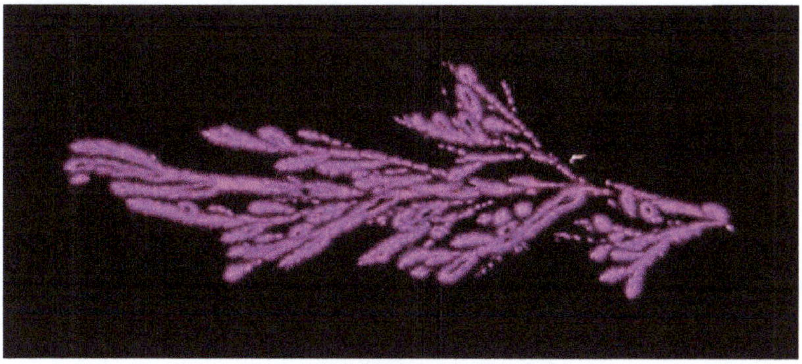

Figure 6-8

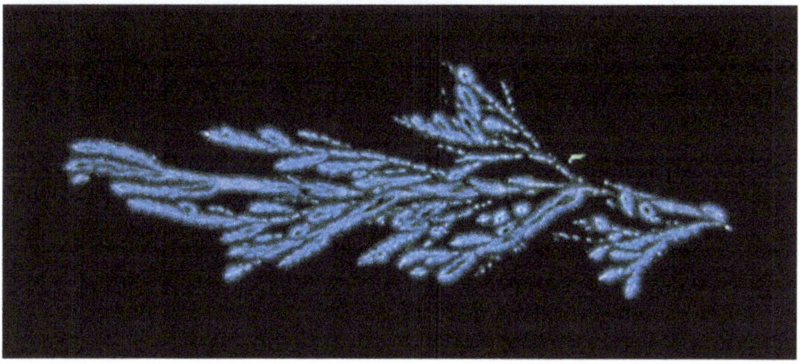

Figure 6-9

Going Further:

There are additional methods one may use to capture kirlian photographs using the transparent discharge plate. The long exposure and light gathering ability of your camera is important. Software, equipment and techniques may be borrowed from astrophotography.

Chapter 7

Kirlian Art Prints for Fun and Profit

Fortunately for you there are not many photographers pursuing Kirlian photography. This allows you greater latitude for creating fine art Kirlian prints that are acceptable for publication and for sale.

You have Kirlian prints that range in the thousands of dollars like Walter Chappell prints selling for $3200.00 on Amazon.com

https://www.amazon.com/Walter-Chappell-Parsley-Kirlian-Platinum/dp/B019BVLOIU/ref=sr_1_3?ie=UTF8&qid=1466435859&sr=8-3&keywords=fine+art+kirlian

You also have more budget priced "Limited Edition" Kirlian prints for $99.95.

https://www.amazon.com/Kirlian-Maple-Leaf-Limited-Edition/dp/B0092GH8NK/ref=sr_1_cc_2?s=aps&ie=UTF8&qid=1466436146&sr=1-2-catcorr&keywords=kirlian+prints

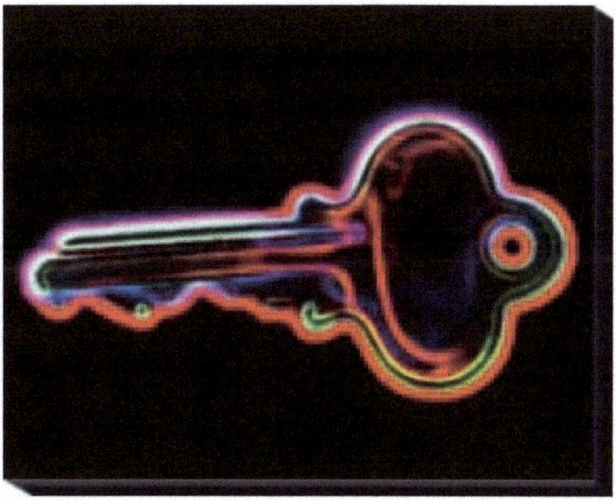

Figure 7-1 -- Canvas Print

And naturally you have fine art prints priced in between.

Printing on Media

Kirlian images may be printed on paper, calendars, cards, canvas, cups, tee shirts, mouse pads, bags and pretty much anything else you can think. Another fertile field are image transfers of Kirlian images onto various surfaces like: metal, wood, plastics and glass.

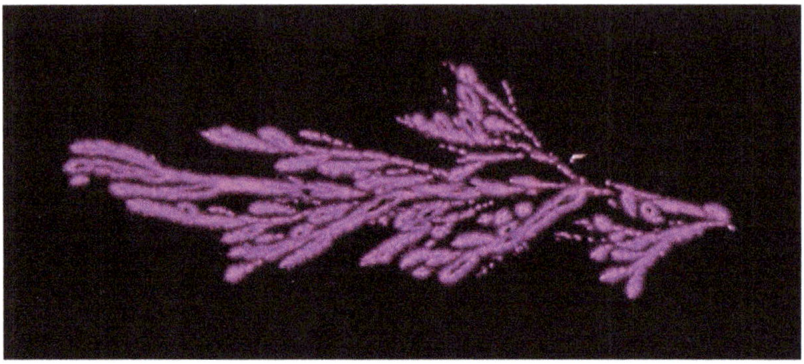

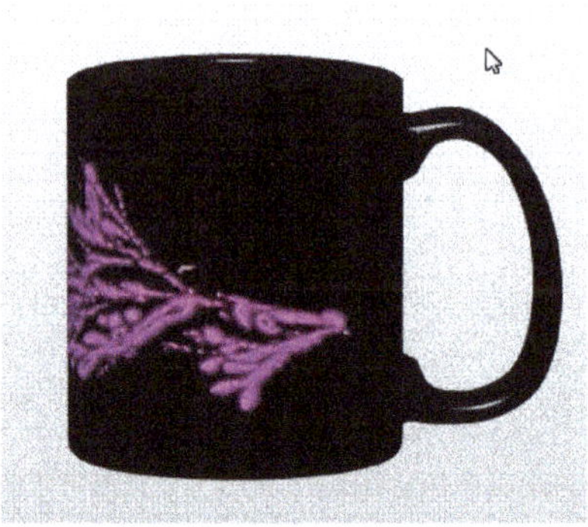

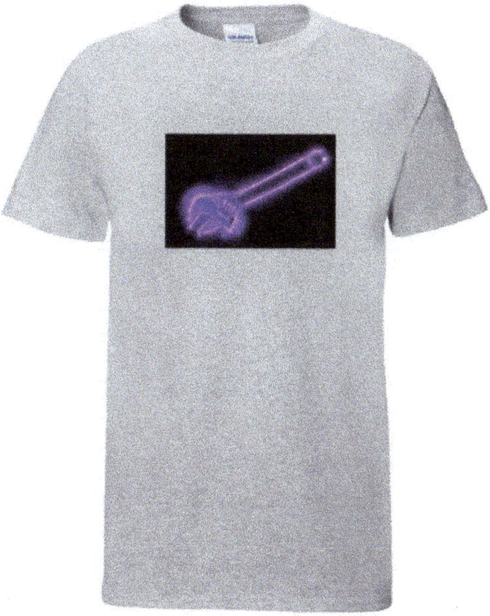

Markets:

You can offer your Kirlian artwork on craft sites like Etsy, major retailers like Amazon.com, Art Galleries and studios, local craft shows, and of course your own website.

Resources:

There are many companies that print on media, I have included a few below:

https://www.shutterfly.com/

http://www.vistaprint.com/

Gallery

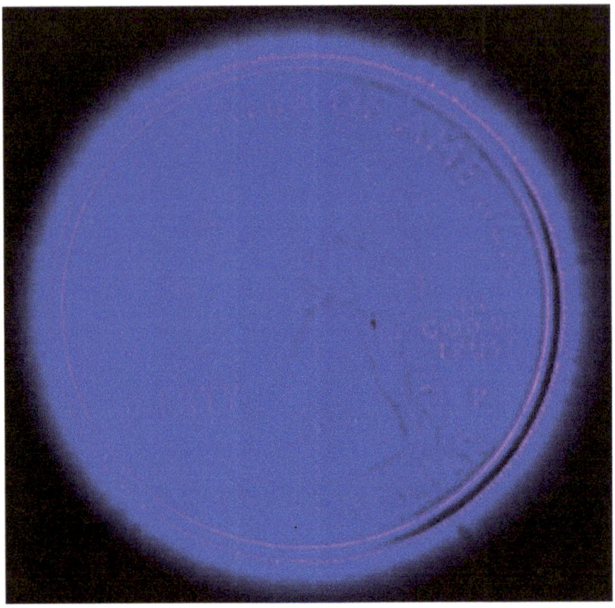

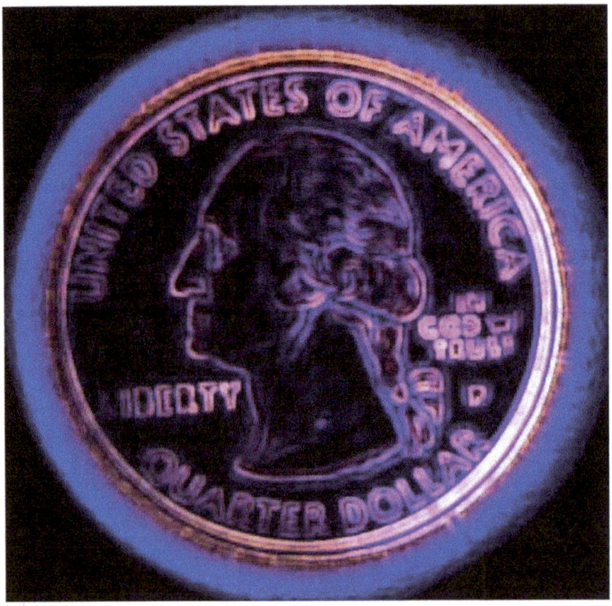

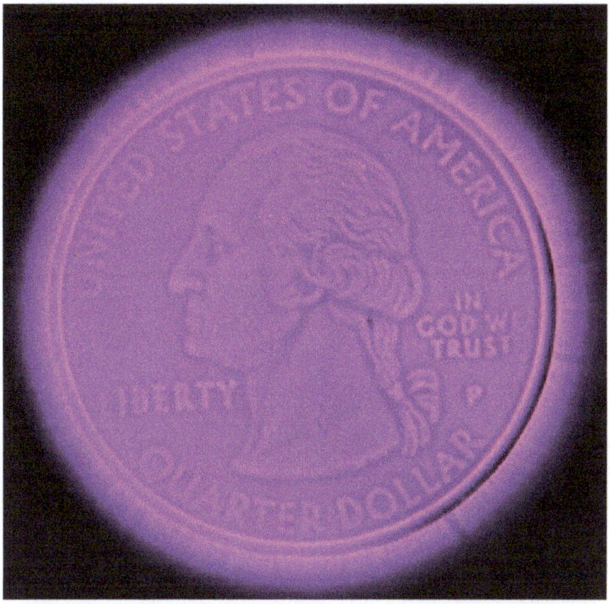
UNITED STATES OF AMERICA
IN
GOD WE
TRUST
LIBERTY
P
QUARTER DOLLAR

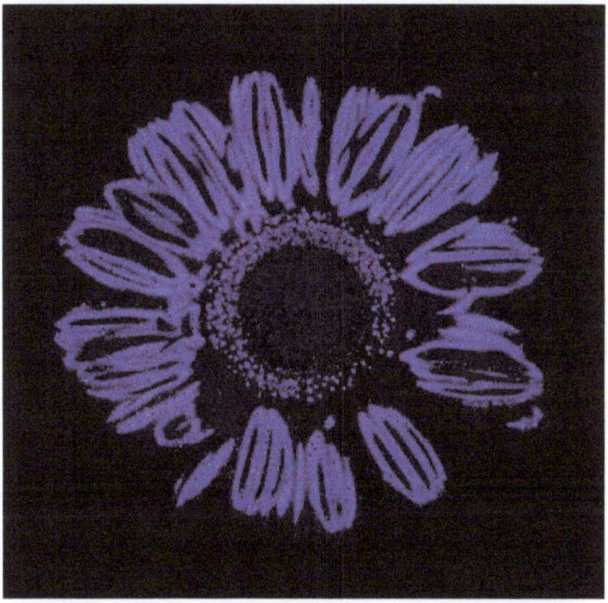

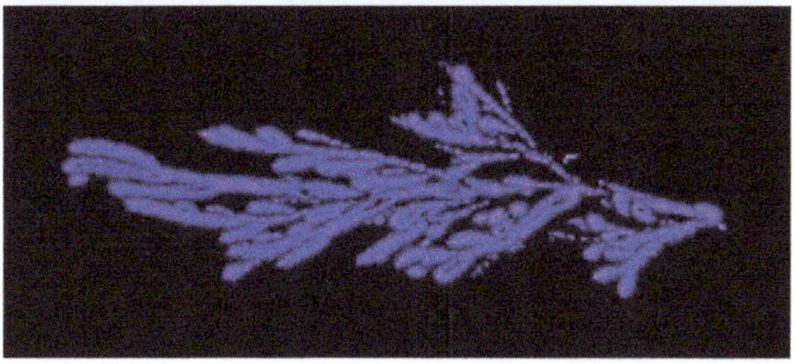

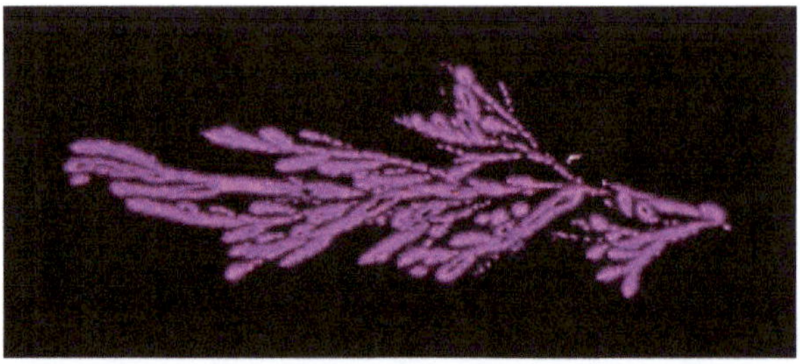

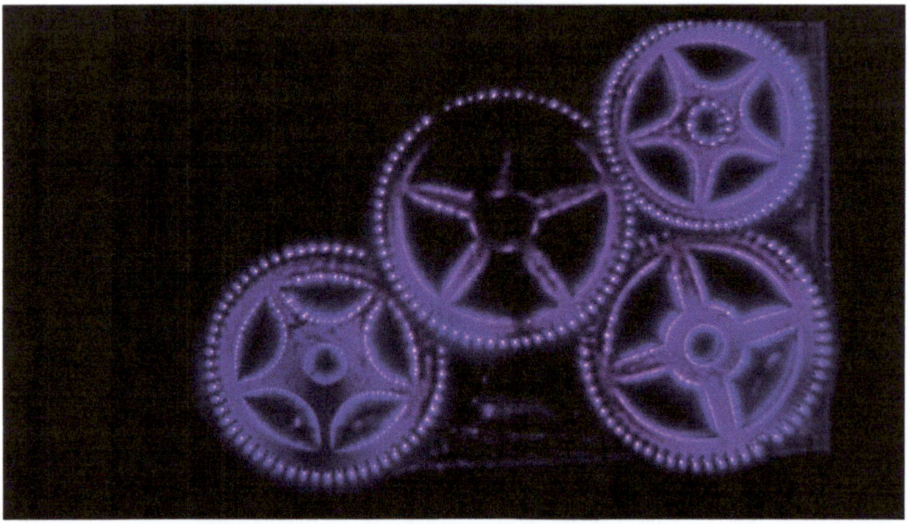

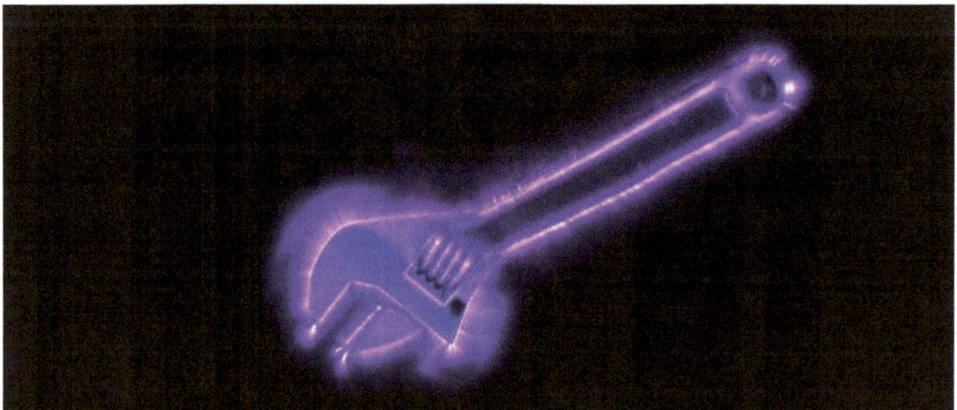

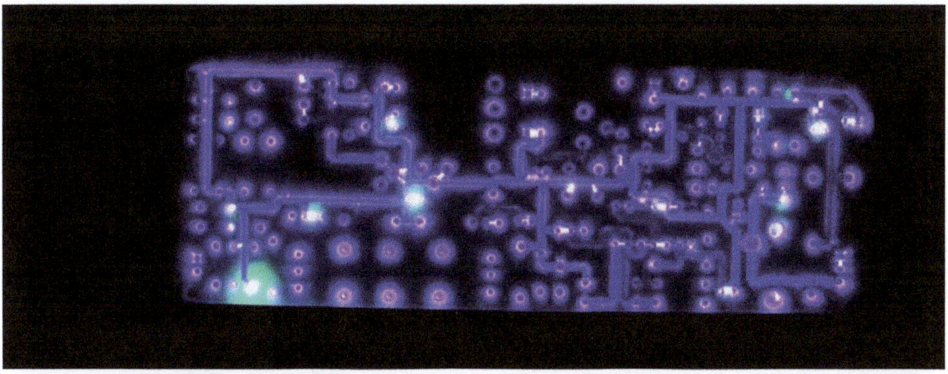

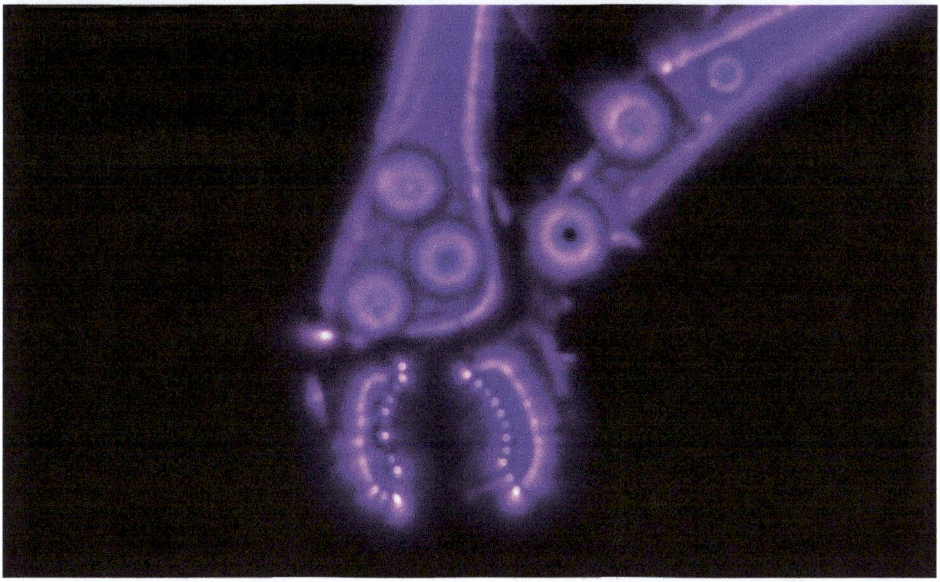

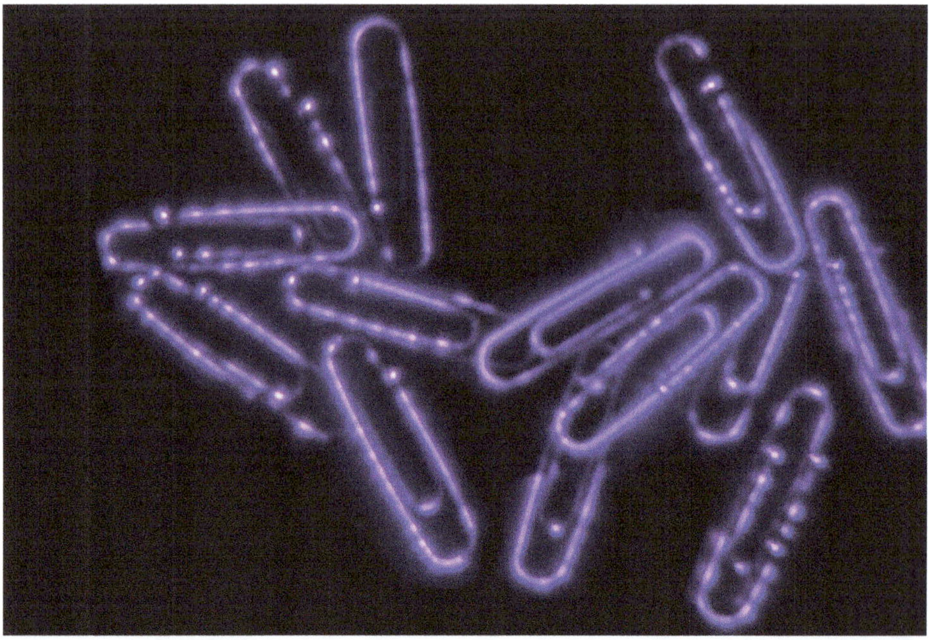

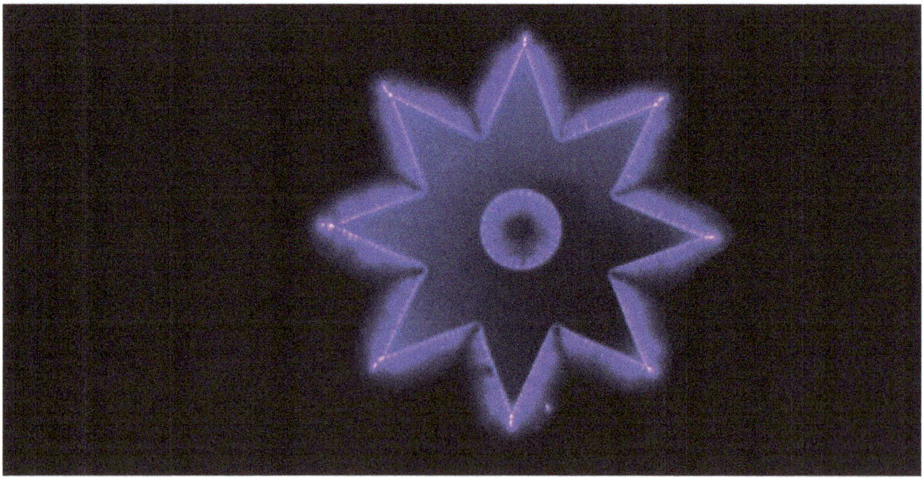

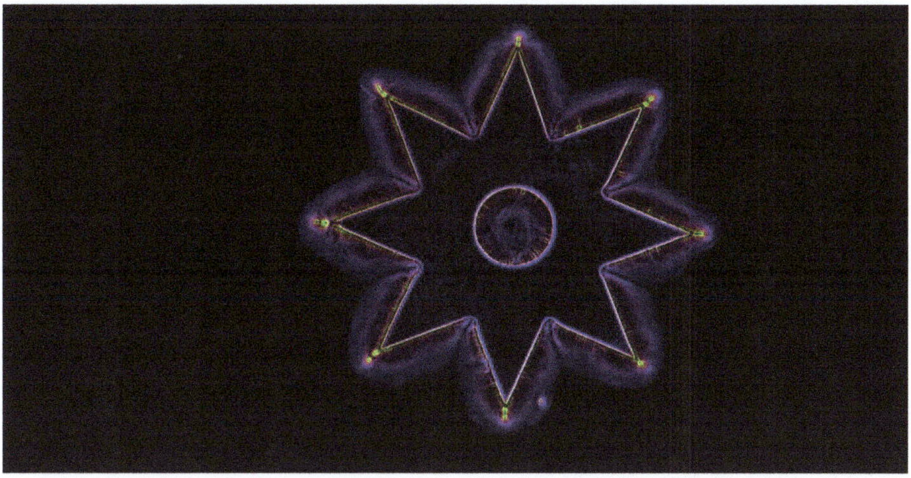

Learning More about Kirlian Photography

If you want to learn more about Kirlian photography, I recommend my other book Kirlian Photography Fact and Fiction.

This book links to private documentary videos

www.ingramcontent.com/pod-product-compliance
Lightning Source LLC
LaVergne TN
LVHW051953260426
837140LV00022B/53
9781623850043